RADON

Edited by **Feriz Adrovic**

Radon
http://dx.doi.org/10.5772/66540
Edited by Feriz Adrovic

Contributors

Ayman Mohamed Abdalla, A. M. Ismail, Tayseer Ibrahim Al-Naggar, Jan Kozempel, Abd Elmoniem Ahmed Elzain, Akbar Abbasi, Zejnil Tresnjo, Jasmin Adrovic, Ema Hankic, Feriz Adrovic, Diana Linhares, Patricia Garcia, Armindo Rodrigues, Rakhmetkazhy Bersimbaev, Olga Bulgakova

Published by InTech
Janeza Trdine 9, 51000 Rijeka, Croatia

Publishing Process Manager Mirena Calmic
Technical Editor SPi Global
Cover InTech Design team

First published November, 2017
Printed in Croatia
Legal deposit, Croatia: National and University Library in Zagreb

Additional hard copies can be obtained from orders@intechopen.com

Radon, Edited by Feriz Adrovic
p. cm.
Print ISBN 978-953-51-3655-2
Online ISBN 978-953-51-3656-9

For EU product safety concerns:
IN TECH d.o.o., Prolaz Marije Krucifikse Kozulić 3, 51000 Rijeka, Croatia,
info@intechopen.com or visit our website at intechopen.com.

Meet the editor

Feriz Adrović, PhD, is a full professor at the Department of Physics, Faculty of Science, University of Tuzla, Bosnia and Herzegovina. His main areas of interest are applied nuclear physics, radon, gamma spectrometry, radiological physics, and radiation protection. His current interest is biomonitoring. Dr. Adrović is the founder of the prestigious scientific research center for detection and dosimetry of the University of Tuzla, Bosnia and Herzegovina. For many years, he has been working on the development of radon detection dosimeters. He was a mentor of 8 doctoral theses and 16 master's theses. He has written 2 university textbooks and 2 monographs; he is also the editor and author of the book Gamma Radiation and a reviewer in 11 journals. He has published more than 100 scientific papers in domestic and international journals. He has participated in a large number of scientific congresses and conferences.

Contents

Preface IX

Chapter 1 **Introductory Chapter: Radon Phenomenon 1**
Feriz Adrović

Chapter 2 **Radon Nuclides and Radon Generators 7**
Ján Kozempel

Chapter 3 **Radon Monitoring in the Environment 25**
Abd Elmoniem Ahmed Elzain

Chapter 4 **Levels of Radon and Granite Building Materials 47**
Akbar Abbasi

Chapter 5 **Levels of Radon Activity Concentration and Gamma Dose Rate in Air of Coal Mines in Bosnia and Herzegovina 59**
Zejnil Tresnjo, Jasmin Adrovic and Ema Hankic

Chapter 6 **Radon Exposure and Human Health: What Happens in Volcanic Environments? 79**
Diana Linhares, Patrícia Garcia and Armindo Rodrigues

Chapter 7 **Residential Radon Exposure and Lung Cancer Risk in Kazakhstan 93**
Rakhmetkazhy Bersimbaev and Olga Bulgakova

Chapter 8 **Radon Calibration System 125**
Ayman M. Abdalla, Ahmed M. Ismail and Tayseer I. Al-Naggar

Preface

My intention in preparing this book was to present the most recent research on radioactive gas radon. For many years, I've been dealing with various radon-related problems, so there is a need to show the latest and most reliable knowledge of this gas that poses a risk to human health. A better opportunity could not be found in the realization of this idea since publishing the book in the world's leading Open Access publisher InTech, Rijeka.

Radon and its descendants of radioactive disintegration are permanently present in the environment, and realistic facts need to be considered in the present and in the future when considering their impact on human health. This effect represents a health radiological risk, which is in most cases small in relation to other routine risks associated with radioactive material or exposure to radioactive radiation due to nuclear accidents. Throughout the world, epidemiological studies of large populations have been carried out to detect the causal relationship between radon and radon progeny exposure and the appearance of lung cancer. In most of these studies, a positive correlation was found between increased number of lung cancers and increased radon concentration in working and living rooms. Radon was identified as a public health problem.

The book Radon presents segments of modern research from a wide range of issues related to radioactive radon gas. In these chapters of the book, physical and chemical properties of radon and radon progeny; concentration, emanation, and transport of radon in ambient environments; detection of radon and radon progeny in different environments; passive and active radon measurement techniques; and calibration of a dosimeter for detection of radon are shown. Although the book contains a relatively small number of chapters, the most important problems in this broad field of research are considered in a quality and concise manner. I thank the authors of the chapters and the bookmakers for their hard work and professionalism. They have directly contributed to the quality of the book.

This book will be of great importance to scientists from a wide range of research area where radon phenomena are applied, and will be useful as well for those who are beginners in this topic. Due to the health importance of radon gas, the content of this book will be interest to a wider audience as well.

Prof. Feriz Adrovic
Faculty of Science
University of Tuzla
Bosnia and Herzegovina

Chapter 1

Introductory Chapter: Radon Phenomenon

Feriz Adrović

Additional information is available at the end of the chapter

http://dx.doi.org/10.5772/intechopen.70170

Live beings on earth have always been exposed to radiation from nature and more recently from artificial sources of radiation. The main components of radiation from nature are cosmic rays, terrestrial gamma ray, ingestion, and inhalation of natural radionuclides. On normal occasions, terrestrial sources are responsible for most of human exposure to natural radiation. These are, above all, radionuclides that are members of the three natural radioactive series: uranium-radium (^{238}U), uranium-actinium (^{235}U), and thorium series (^{232}Th). Usually, the radionuclides on the beginning of these three radioactive chains are called primordial or primary natural radionuclides. All members of these series are genetically linked and are the result of the successive decay of the first member of the series, which explains the law of radioactive decay. Natural radioactivity is the occurrence of atomic core decomposition that exists in nature, without external influences, at which alpha particles (helium nucleus), beta particles (electrons and positron), and cosmic rays (photons) are emitted.

Researches of natural radioactivity include a wide range of science disciplines, starting with physics as a fundamental science, which explains the phenomenology of radioactive disintegration and measurements of activity of radioactive elements, and geological disciplines such as geochemistry, radionuclide prospecting, and physical chemistry, which explains the mobilization and spreading of radionuclides in nature. In the environment, radionuclides of natural origin, at most locations, are mainly present in low concentrations. However, these radionuclides through various geochemical processes can be significantly concentrated regionally or locally, leading to even their mineralization and/or forming mineral deposits, such as in the case for isotopes U and Th [1, 2]. The radium follows the secondary uranium concentration but often forms separate aureole dispersion due to certain differences in geochemical characteristics. Radium, as a dissolvable cation, is migrative under conditions where the uranium is immobile and in the water or land can be to a greater or lesser degree separated from uranium and locally separately concentrated.

Natural radionuclides of terrestrial origin enter into the environment in various ways and from different sources: natural, geochemical (mineral deposits of phosphate, lead and zinc and other raw materials, various rocks), and anthropogenic sources (mining and processing, production and use of mineral fertilizer, radioactive waste dumps, ash emissions, ash and slag dumps from thermal power plants). Three natural radionuclides that are representing the greatest environmental risks, uranium, radium, and radon, appear in different parts of the environment: in soil, underground, surface waters, and air, either together or separated [3]. These natural terrestrial radionuclides are the most common and most important sources of ionizing radiation in the environment, among which the radon isotope ^{222}Rn occupies a central place, both in terms of total irradiation of the population and in terms of local high-radiation doses. Ionizing radiation is one of the most dangerous health risks in the environment.

Radon is a subject of intensive research around the world. Problems related to this radioactive gas are very complex and are representing wide field for scientific research. Radon is of special health importance for humans, and this is a motive, among other things, why there is a permanent interest in it. The study of the biological effects of radon on man is a very demanding scientific task. The human body is a very complex system with many organs of different chemical properties, functions, and radiosensitivity, so the reliable estimation of exposure risk, the development of adequate protection from radon, requires a multidisciplinary approach to these problems.

Radon as an inert gas, almost without forming any compounds in nature, is a global natural phenomenon. Radon isotopes, ^{222}Rn, ^{220}Rn, and ^{219}Rn, with a half-life of 3.825 days, 55.6 s, and 3.96 s, respectively, are the central members of each natural radioactive chain. Their immediate predecessors are radium isotopes, whose atoms disintegrate into radon via alpha decay. The isotope ^{219}Rn is insignificant due to the small amount of ^{235}U, while the isotope ^{220}Rn is of less importance due to its short half life. Because of the longest life time and isotope abundance, the highest importance of natural radon isotopes is ^{222}Rn, and the term radon refers to this isotope. In its decay chain, the radon ^{222}Rn gas is transformed into a stable lead ^{206}Pb through the short-living descendants of disintegration, emitting five alpha particles of energy up to 7.7 MeV, a beta radiation of up to 2.8 MeV, and a photon of gamma radiation of energy up to 2.4 MeV.

Humans are constantly exposed to radiation from many sources from nature and in the last 100 years from artificial sources. Over 80% of global exposure is from natural sources, and about 20% comes from artificial sources—mostly from radiation used in medicine. Natural radioactive sources contribute with 2.4 mSv to an average annual effective dose, and it varies from about 1 to more than 10 mSv, depending on where people live. The average annual effective dose of ^{220}Rn inhalation and its progeny is 70 μSv and by ^{222}Rn inhalation and its progeny is about 1.2 mSv [4]. For this reason, the problem of the radon level in the environment is treated with special care.

The high mortality rate of miners in central Europe was noted in the sixteenth century, but only at the end of the nineteenth century, it was concluded that the lung cancer was the cause of this phenomenon. The measurements of radon activity concentrations in the Schneeberg mines and Jachymov mines (ranging from 70 to 120 kBq/m^3), in the 1930s and 1940s, showed that the lung cancer of miners in these mines was caused by high radioactive radon gas levels. Finally, after almost 500 years, the mysterious deaths in these mines, called "Bergsucht" or "mountain sickness," have been puzzled out. Measurements of radon activity concentrations in the mines of leading industrial countries in the world were followed. These measurements confirmed the radon damage to the human health, and the risk awareness of lung cancer among miners was slowly but persistently matured. This had finally encouraged the motivation for a professional attitude toward radioactive radon gas, providing new and more effective protection guidelines for underground mines, especially uranium mines.

Based on the significant experience gained through numerous researches of radon in uranium mines, as in other mines, radon and radon progeny have been identified as causes of lung cancer [5–8]. Epidemiological studies on levels of radon activity in underground mines have become the main basis for estimating quantitative risk for lung cancer associated with exposure to radon and radon decay products. Surprisingly, the space where a man lives up to 70% of his time in life was not a priority in the initial stages of radon concentration research. Measurements of radon activity concentrations in indoor enclosures were made for the first time in 225 houses in Sweden, whose results were officially published in 1956. The initiative for these studies, given by Rolf Sievert, was proved to be justified because in a number of houses that were made of porous concrete with increased content of ^{226}Ra; the radon concentration was noticeably higher than in the other houses. These researches did not attract the attention of the international experts, because they were considered to have local significance and therefore were forgotten.

The return of interest for the research for this radioactive gas was only 20 years later, again in Sweden, due to energy saving during oil crisis. The Swedish government's commission points out that the radon problem is a common problem and that it exists outside of Sweden. The United Nations Scientific Committee on the Effects of Atomic Radiation (UNSCEAR) quickly reacted, and in the 1982 report [5], for the first time, the radon contribution to the average radiation exposure in natural conditions was taken into account. Similarly, in 1983, International Commission on Radiological Protection (ICRP), in its publication ICRP-39, the Principles for Limiting Exposure of the Public to Natural Sources of Radiation, issued the first recommendations for this specific natural radionuclide [9].

Researches in Sweden gave motive for systematic measurements of radon concentration in residential and workplaces around the world. Numerous studies have shown that the concentration of radon activity in homes depends on a large number of factors, such as design and type of home, local geology, soil permeability, building material and building ventilation, etc. All of these factors are variable, even in a small area, and therefore, radon concentrations can be significantly different even between neighboring dwellings and houses. Concentrations of indoor radon in buildings, homes, schools, and offices at investigated locations in the world range from 10 to 10,000 Bq/m^3 [10]. Fortunately, the number of objects with such high concentrations of radon is very small. Certain mentioned factors give a dominant contribution to the concentration levels of the indoor radon activity.

The health effect that radon has on the population has ultimately become the primary cause for the interest of the scientific audience for radon. In addition, with scientific and expert research on the radon problem, numerous possibilities of the physicochemical property application of this radioactive gas have been revealed. This is primarily related to the prospect of uranium and then for geothermal investigations, earthquake prediction, and volcanic eruptions.

Measuring the concentration of radon activity in groundwater can identify different cracks as well as geothermal sources that lie deep beneath the earth's surface. The presence of geological cracks enables radon to increasingly exhalates from the soil through transport processes, which are related to the deep bursts of earth [11].

A significant role in the total population irradiation has the content of radon in drinking water but also water used for other purposes. This primarily applies to water used in swimming pools, as well as baths and tubs for inhalation in recreational and medical centers. Radon from water contributes to the total inhalation risk associated with indoor air radon. The most extensive research on radon water risk is presented in the report of the US Scientific Committee on Risk Assessment of Radon in Drinking Water. Underground water often travels through rocks that contain radium, which in the process of disintegration releases radon into water [12]. The radiation dose for radon that enters the body with drinking water depends on its concentration activity in the water, metabolism, and radon kinetics in the body. Radon dissolves in water, body, and blood. Radiation emitted by radon and its products of radioactive decay displays sensitive stomach cells as well as other organs after it is being absorbed into the bloodstream. It is easily absorbed from the gastrointestinal tract and distributed among the tissues, partly due to a large coefficient of emanation and partly due to its relative solubility in blood and tissue.

Many studies have crystallized evidence that cell exposure to high radiation, such as alpha particles, can trigger a series of molecular and cellular events, DNA breaks, inaccurate repairs, apoptosis, gene mutations, chromosome changes, and genetic instability, to culminate in the end in the development of lung cancer and cancers of the other branches [13–17]. Radon was identified as a public health problem [18]. The complexity of observing radon dosimetry on biotics systems is also related to the fact that radon and its short- and long-lived products are alpha, beta, and gamma emitters. These various forms of ionizing radiation, with various energies and physical-chemical tendencies, give the whole range of biological effects on the way of transporting radon through tissues of the organisms.

On the other hand, treatment with radon is one of the oldest therapies used by humans, and it is in use today [19]. Radon treatment involves inhalation from natural sources (ore mine) or bathing in the water that contains radon, while there is less use of radon-rich drinking water [20, 21]. The first results of radon-rich water activity were observed in patients with acute and chronic rheumatism, motor regeneration, posttraumatic and postsurgical conditions, and other diseases of the organism [20, 22]. Particularly, positive radon activity was observed in the cardiovascular system, certain radon levels lead to cardiac contractions, and arterial pressure drops. These are special "hormesis" [23].

According to US Environmental Protection Agency (EPA) estimates, radon is the cause of the number one lung cancer among nonsmokers and the second leading cause of lung cancer after smoking. Although radon is one of the most widely investigated human carcinogens, which is proved the most in epidemiological researches on lung cancer in underground miners, the effect of small doses has not yet been sufficiently resolved. By developing new dosimetry models and new measuring systems that are adjusted to realistic measuring conditions, knowledge of the biological effects of radon and its degradation products will be complemented.

One of the missions of the book *Radon* is to keep the radon problem always in the sphere of human interest.

Author details

Feriz Adrović

Address all correspondence to: adrovicferiz@yahoo.com

Department of Physics, Faculty of Science, University of Tuzla, Bosnia and Herzegovina

References

[1] Faure G. Principles of isotope geology. 2nd ed. New York: Wiley, ©1986

[2] Fritz P, Fontes JC, editors. Handbook of environmental isotope geochemistry. Vol. 1. Elsevier; Amsterdam,1980

[3] Gržetić I, Jelenković R. Prirodni radioaktivni elementi, geološko porijeklo i oblici pojavljivanja i migracije. Monografija: Jonizujuća zračenja iz prirode, JDZZ, Beograd; 1995. pp. 1-39

[4] UNSCEAR. Report to the General Assembly, with Scientific Annexes, Sources and Effects of Ionizing Radiation, United Nations Scientific Committee on the Effects of Atomic Radiation. New York: United Nations; 1993

[5] UNSCEAR. Report to the General Assembly, with Scientific Annexes, Sources and Effects of Ionizing Radiation, United Nations Scientific Committee on the Effects of Atomic Radiation. New York: United Nations; 1982

[6] ICRP. Protection Against Radon-222 at Home and at Work. ICRP Publication 65. 1993

[7] ICRP. Human Respiratory Tract Model for Radiological Protection. ICRP Publication 66. 1994

[8] UNSCEAR. Sources and Effects of Ionising Radiation. Report of the United Nations Scientific Committee on the Effects of Atomic Radiation; 2000

[9] ICRP. Principles for Limiting Exposure of the Public to Natural Sources of Radiation. ICRP Publication 39. 1984

[10] WHO, Radon and Health, June 2016

[11] Fleischer RL. Radon: Overview of properties, origin and transport. In: Durrani SA, Ilic R, editors. Measurements by Etched Track Detectors. Singapore: World Scientific; 1997. pp. 3-18

[12] National Research Council. Risk Assessment of Radon in Drinking Water. 1999

[13] International Atomic Energy Agency (IAEA). Radiation Biology: A Handbook for Teachers and Students. 2010

[14] Kronenberg A. Mutation inductions in human lymphoid cells by energetic heavy ions. Advances in Space Research. 1994;**14**(10):339-346

[15] Anderson R, Marsden SJ, Wright EG, Goodhead DT, Griffin CS. Complex chromosome aberrations in peripheral blood lymphocytes as a potential biomarker of exposure to high-LET alpha particles. International Journal of Radiation Biology. 2000;**76**:31-40

[16] Maffei F, Angelini S, Cantelli FG, Lodi V, Mattioli S, Hrelia P. Micronuclei frequencies in hospital workers occupationally exposed to low levels of ionizing radiation: Influence of smoking status and other factors. Mutagenesis. 2008;**17**:405-409

[17] WHO Handbook on Indoor Radon: A Public Health Perspective. World Health Organization; 2009

[18] Environmental Protection Agency (EPA). A Citizen's Guide to Radon: The Guide to Protecting Yourself and Your Family From Radon U.S. 2005

[19] Falkenbach A, Wolter N. Radonthermalstollen-kur zur Behandlung des Morbius Bechterew, Research in Complementary Medicine. 1997;277-283

[20] Parker D. Letter to Sponsors of the Free Enterprise Radon Health Mine, Boulder, MT, regarding trip to Moscow, Austria, West Germany, and France. 1987

[21] Becker K. One century of radon therapy, International Journal of Low Radiation, 2004;**1**:334-357

[22] Cohen B. Test of the linear-no-threshold theory of radiation carcinogenesis for inhaled radon decay products. Health Physics. 1995;**68**:157-174

[23] Macklis R, Beresford B. Radiation hormesis. Journal of Nuclear Medicine. 1991;**32**:350-359

Chapter 2

Radon Nuclides and Radon Generators

Ján Kozempel

Additional information is available at the end of the chapter

http://dx.doi.org/10.5772/intechopen.69901

Abstract

The radon element is the heaviest and the only naturally occurring radioactive noble gas. As a member of uranium and thorium decay chains, it is formed instantaneously and belongs to the naturally occurring radioactive materials (*NORM*). The long-lived radon isotope, the ^{222}Rn, is radiobiologically the most important one. It is present in subsoil and groundwater and permeates to the surface, where it may become health risk during the long-term inhalation. Proper testing of drinking water and building materials is also required to monitor radon concentrations below legal limits. Thus, the need of radon determination as well as the preparation of its isotopes arises for its use as a calibration source for the environmental and workplace monitoring in the *NORM* as well as other industries. Further, the radon isotopes currently appear in various research fields, including radionuclide progeny preparation and their use is experiencing renaissance. An overview of radon characteristics, its physical and chemical properties, as well as radon isotope preparation methods including the radionuclide generators and their use is given here. Radon isotope use for tracing, medical, geochemical and other purposes is also discussed.

Keywords: radon, thoron, actinon, generator, decay chain, radioactive deposits, Rn, noble gas

1. Historical introduction

Soon after the discovery of radium by Curie et al. [1], the radioactivity of thorium by Schmidt [2] and the discovery of actinium by Debierne [3], it was found that all of these radioactive elements activate their surroundings and emit formerly unknown radioactive gases. Curie and Curie discovered that objects exposed to radium samples got activated and the half-life of the gained radioactivity was approximately 1 month [4], even though clear interpretation of this discovery remained unknown. Similarly, radioactive gas evolution was observed in samples of thorium and uranium [5–9]. Rutherford and Dorn interpreted the emanated activity from thorium and radium compounds as new chemical substances. Rutherford and Owens reported that the gas released from thorium had the half-life of about 1 min. Radioactive gas release from actinium samples was reported by Debierne soon after [10]. All the three gases (originating from radium, thorium and actinium) were found as chemically inert and were classified as noble gases [11]. Further studies on emanation gases lead to better understanding of their physical and chemical properties as well as their nature, for example, their condensation at low temperatures [12]. Many experiments were performed in order to determine the atomic mass of the emanation. In 1910, among others, Ramsay and Gray collected radium emanation gas to measure its density and determined the value of atomic mass to be around the value of 220 [13]. Later on, the formerly known and so-called radium, thorium and actinium emanations were named as radon, thoron and actinon, respectively [14]. The radon nuclides were originally important subjects of basic science [15]. After the determination and precise measurements of all their physical and chemical properties, various applications of radon appeared such as its use in medicine, geology, tracing applications, radionuclide production and research.

2. Radon nuclides and their characteristics

Radon (the element) is the heaviest known noble gas. Radon has only radioactive nuclides. In total, some 40 isotopes of radon are known up today. Three most common and naturally abundant isotopes of radon are ^{222}Rn, ^{220}Rn and ^{219}Rn, originating from the decay series of ^{238}U, ^{232}Th and ^{235}U, respectively. Their decay chains are shown in **Figures 1–3**, and the nuclear data are summarized in **Table 1**. Nuclear data were taken from Refs. [16, 17].

Some selected physical and chemical properties of radon element are summarized in **Table 2**. Data were taken from Ref. [18]. Radon belongs to the noble gas group that predicts its parameters.

As it could be clearly seen from **Figures 1–3**, the daughter nuclides of radon generate radioactive progeny of Po, At, Pb, Bi, Hg and Tl. The decay of Radon in air (gas phase) results in the spread of radioactive aerosols and their deposits mainly in case of ^{222}Rn may become hazardous (long-lived deposits). Radiotoxicity of radon is thus an important factor, since it may be released from natural sources (e.g., building materials, subsoil, drinking water and mine air) and inhaled.

Following the natural decay series (^{238}U, ^{232}Th and ^{235}U), the nuclides of ^{222}Rn, ^{220}Rn and ^{219}Rn are instantaneously formed progeny in the environment and their mother nuclei samples. These isotopes may be, however, produced in various nuclear reactions, for example, the ^{219}Rn was co-produced in the experiment of chemical characterization of the element copernicium (^{283}112) in the nuclear fusion of ^{48}Ca with ^{242}Pu, where the ^{219}Rn served as a calibration nuclide in the thermochromatographic separations of the element 112 in the *COLD* detector [19]. Other exotic isotopes of radon may be produced in various nuclear reactions. For example, the lightest-known neutron deficient isotopes of radon, ^{193}Rn and ^{194}Rn, were produced in the complete fusion reaction of ^{52}Cr pulsed ion beam with a ^{144}Sm target [20]. The ^{209}Rn and ^{210}Rn isotopes were determined in the decay chains of ^{213}Ra and ^{214}Ra, respectively, prepared

http://dx.doi.org/10.5772/intechopen.69901

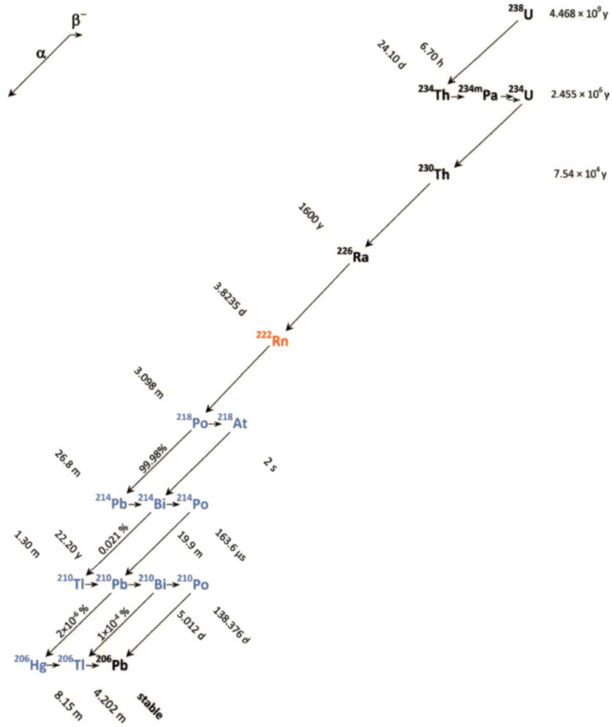

Figure 1. ^{238}U decay chain.

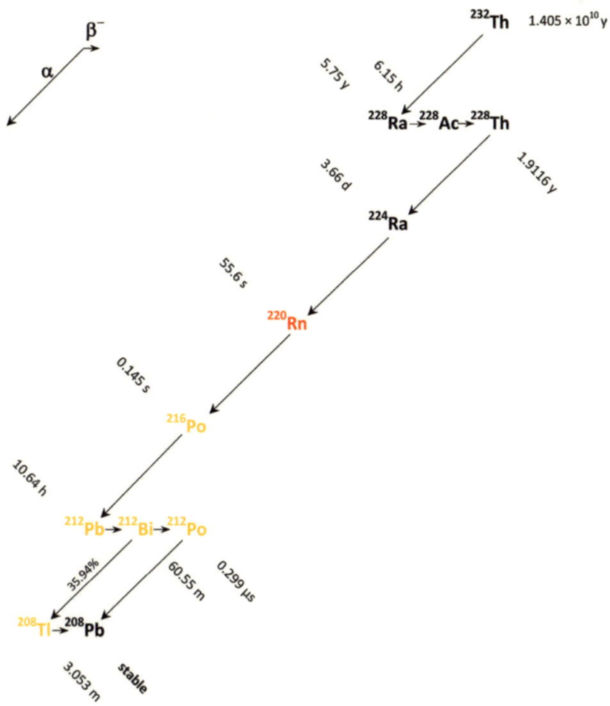

Figure 2. ^{235}U decay chain.

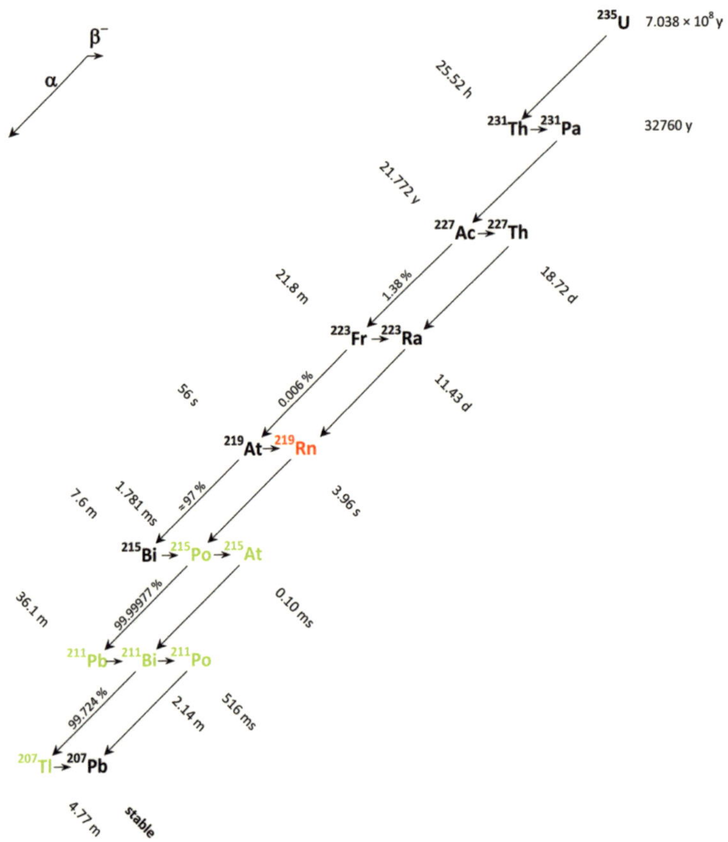

Figure 3. ^{232}Th decay chain.

Historical name	Radon	Thoron	Actinon
Nuclide	^{222}Rn	^{220}Rn	^{219}Rn
Half-life	3.8253 d	55.6 s	3.96 s
Generator nuclide	^{226}Ra (^{238}U) 1600 y	^{232}Th (^{228}Th) 1.405 × 10^{10} y	^{227}Ac (^{235}U/^{231}Pa) 21.772 y
Main α radiations E (MeV) (intensity (%))	5.4895 (99.92) 4.987 (0.078) 4.827 (~0.0005)	6.288 (99.886) 5.747 (0.114)	6.819 (79.4) 6.553 (12.9) 6.425 (7.5) 6.529 (0.12) 6.312 (0.054) 6.159 (0.0174)
Main γ radiations E (keV) (intensity (%))	511 (0.076)	549.8 (0.114)	271.2 (10.8) 401.8 (6.37) 130.6 (0.119) 293.5 (0.073) 517.6 (0.0443) 221.5 (0.030)

Table 1. Main isotopes of radon.

http://dx.doi.org/10.5772/intechopen.69901

Electron shell configuration	[Xe] $4f^{14}$ $5d^{10}$ $6s^2$ $6p^6$	
Oxidation numbers	2, 3, 4	
Ionization potential (eV)	1st	10.748
	2nd	21.4
	3rd	29.4
	4th	44
Pauling electronegativity	2.0	
Atomic radius (nm)	218	
Molar enthalpy of fusion at (kJ mol^{-1})	12.26	
Molar enthalpy of vaporization (kJ mol^{-1})	74.30	
Surface energy (mJ m^{-2})	29	
Melting point (°C)	−71	
Boiling point (°C)	−61.8	
Solubility of radon at standard pressure in water (cm^3/100 g H_2O)	0°C	51.0
	50°C	13.0
Density (g cm^{-3})	0°C (g)	0.00973
	−61.8°C (l)	4.4
Dynamic viscosity coefficient (mPa s)	0.0213	
Thermal conductivity coefficient at 27°C (W m^{-1} K^{-1})	0.00364	

Table 2. Selected physical and chemical properties of radon.

via $^{170}Er(^{50}Ti,3n)^{217}Th \rightarrow {}^{213}Ra$ and $^{170}Er(^{48}Ca,xn)^{218-x}Ra$ reactions [21]. The ^{211}Rn was prepared as a decay product of the francium beams produced by the spallation of actinide targets (U and Th) [22]. The ^{211}Rn was also prepared by the $^{209}Bi(^{7}Li,5n)^{211}Rn$ reaction [23]. Heavier isotopes such as ^{224}Rn, ^{223}Rn and lighter isotope of ^{210}Rn were prepared by the spallation of $^{232}ThO_2$ [24, 25]. The neutron-rich isotopes of ^{227}Rn and ^{228}Rn were also prepared by the spallation of natural ^{232}Th target with 600 MeV proton beam [26]. The heaviest experimentally detected isotopes of radon, the $^{223-229}Rn$, have been determined for the first time, using the *ISOLTRAP* setup at *CERN ISOLDE* experiment [27].

Radon is chemically quite unreactive gas; however, some exotic compounds of radon, for example, the fluorine compounds of radon [28] or RnH^+, $RnOH^+$, $RnOH_2^+$ molecular ions generated in a plasma ion source [29] were reported; other compounds like RnCO [30], HRnCCH [31] and other radon molecules [32] were predicted. Radon adsorption on charcoal is known for a long time and allows its purification from hydrogen, oxygen and nitrogen [33]. Recently, the association of xenon and radon with tris-(triazole ethylamine) cryptophane was studied. High affinity of these noble gases was observed, and the association constants were determined to $K_a = 42{,}000 \pm 2000\ M^{-1}$ and $K_a = 49{,}000 \pm 12{,}000\ M^{-1}$ for xenon and radon, respectively, at 293 K [34].

3. Radon generators

Thanks to the noble gas characteristics of the radon element, its separation from the decay chain or target materials becomes trivial. Radon generators may be then divided into two main groups based on the medium that is used for its final form—gas and liquid apparatuses. Main issue in radon generators is the emanation efficiency, radon physical form and its radionuclidic purity. The radon emanation power or the efficiency coefficient depends directly on the radon source properties. Crystalline and bulky materials exhibit quite low emanation power—typically few per cent of the mother nuclide activity. On the other hand, the emanation efficiency of properly selected and prepared amorphous and porous materials approaches the values of nearly 100%. The simplest generators include simply the mother nuclides enclosed in an evacuated or normal pressure apparatus, allowing the liberation of radon that is to be collected in a gas or liquid phase. Typically, the insoluble salts of mother nuclides trapped in a porous ceramics (or other inert material) or a sandwich of mother nuclide covered by thin separating layer (e.g., foil) were used for the construction of various types of emanators.

Many radium-based inhalation apparatuses as well as drinking or bath water "activators" (**Figure 4**) with the emanation power in the order of 5000–10,000 *Mache* units in 24 h appeared on marker without any regulation (1 *Mache* unit = 3.64 *Eman* = 3.64×10^{-10} Ci/L = 13.4545 Bq/L).

Further, more advanced systems were developed in order to increase the efficiency and purity of the radon gas. For example, the oxygen and ozone may be chemically removed by passing the radon gas through the heated copper wire; the hydrogen by the heated copper

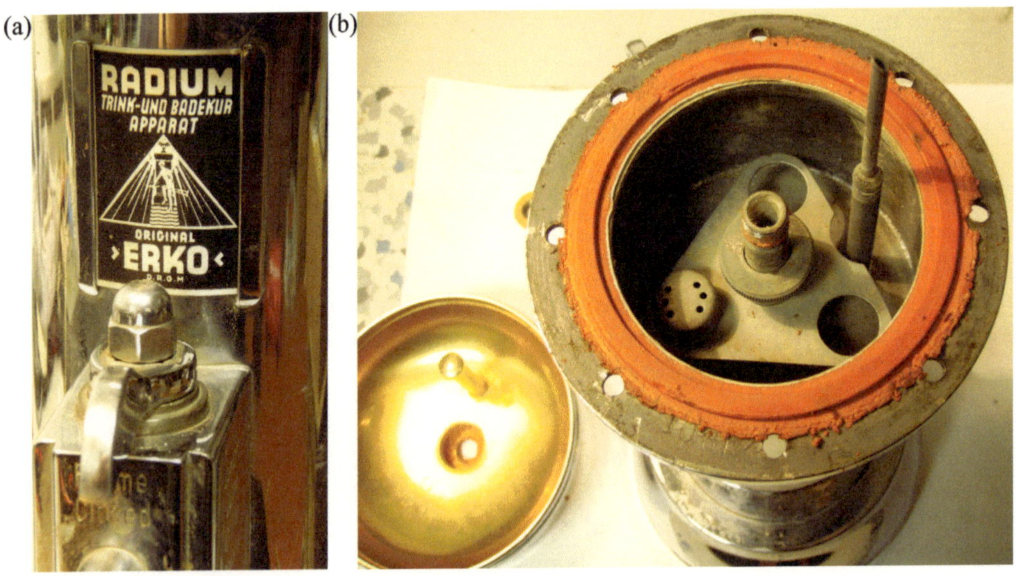

Figure 4. Left—the "ERKO" instrument for the preparation of radium water produced by the Berliner Radium Aktiengesellschaft in 1930s. Right—the same instrument in detail showing a carousel holding one radium capsule with holes to allow free radon emanation (air/water).

oxide; CO_2 by its capture on potassium hydroxide and finally the water on phosphorus pentoxide. Radon of specific activity of 18.5 GBq/mm^3 could be prepared in such way (approx one-third of theoretical volume activity value of pure ^{222}Rn) [35]. Various materials with high emanation power were developed and tested for the construction of radon generators, for example, inorganic porous gels based on heavy and alkali-earth metal hydroxides with hydrated silicic acid were developed [36, 37]. Also the barium stearate powder was reported with the emanating power of >99% for thin layers in air at atmospheric or reduced pressures for actinon [38].

Another important aspect is the source activity metrological standardization. This is necessary firstly for the precise determination of radon contents in various materials and secondly to test the material permeability or retention ability for radon and to verify that these materials meet legal regulations. Various methods for the preparation of calibrated emanation standards were published. Standards containing the solution of $^{226}Ra(NO_3)_2$ absorbed into $CaCO_3$ were prepared, and the emanation coefficient of ^{222}Rn for these standards varied from 0.23 to 0.25 [39]. Accurate and long-term stable sources of defined activity of ^{222}Rn in gas phase were developed for laboratory and field applications (**Figure 5**) [40]. Radon is released from thin layer of a plastic foil with emanation power coefficient approaching 1. The source is constructed as a stainless steel cylinder supplied with the two ball valves on the ends and the two aerosol filters connected on the output aperture of the valves.

Several systems for the preparation of radon in water standard sources were reported. Standard based on an earlier and previously described prototype consisting of polyethylene-encapsulated ^{226}Ra solution source in a small-volume accumulation chamber was used to generate and accurately dispense radium-free ^{222}Rn solutions of known concentration [41]. More recent radon in water standard was developed to get the radon solutions of 300–2000 Bq/L [42] The generator consisted of about 6 L cylindrical vessel with a solid phase ^{222}Rn source with 99.9% air emanation power and an external circuit for solution homogenization. Another radon generator and delivery system was used with 2.9 GBq of radium salt for cell cultures exposure studies [43].

Interesting technique for the radon source preparation that is suitable for use in a low-background liquid scintillation detectors was reported [44]. Radon was concentrated from air to prepare liquid scintillation counting (LSC) sources spiked with activities of 10^6 Bq/m^3. CO_2 and water vapour were removed, and the radon was collected in a cooled charcoal trap. The

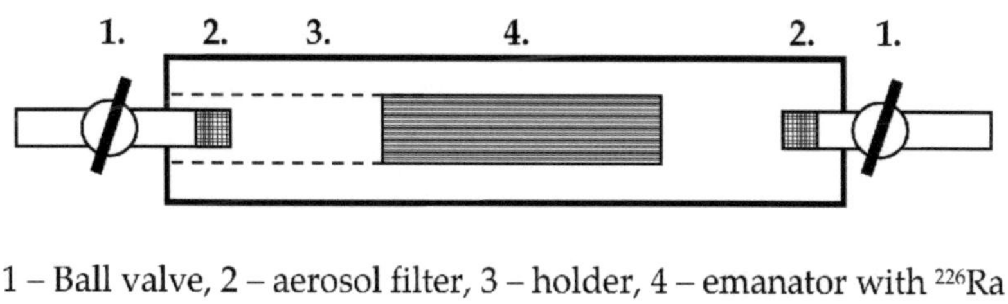

Figure 5. Drawing of radon flow-through generator (air/air).

accumulated radon was desorbed and transferred into a 1,2,4-trimethylbenzene-based scintillator. The sources have been used for the calibration.

Simple laboratory demonstration apparatus for thoron (^{220}Rn) preparation could be constructed, using aged sample of thorium nitrate solution [45]. The solution is enclosed in a bubbler flask connected to a small compressor from where the flowing air displaces the thoron through the valves. Lucas-type scintillation counter [46], ionization chamber or a cloud chamber could be directly connected to the emanation flask through the drying column. Thoron half-life could be easily determined by counting the gas activity enclosed in the detector over few minutes or alternatively the alpha-particle tracks could be visualized in a cloud chamber (**Figure 6**) for educational and demonstration purposes.

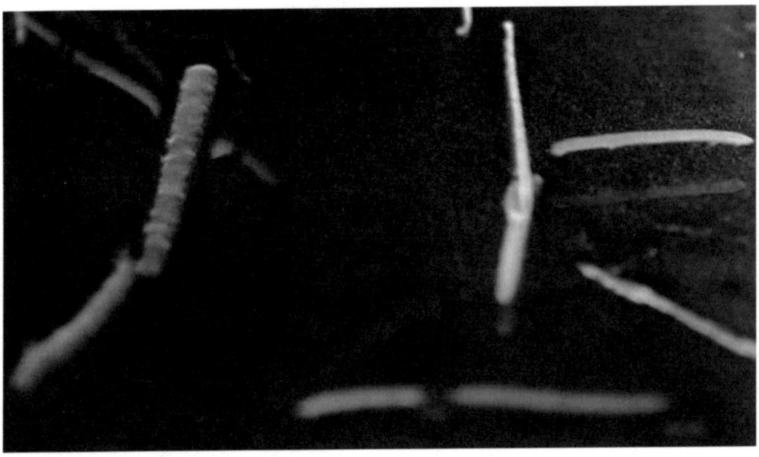

Figure 6. Tracks of alpha particles from thoron (^{220}Rn) decay visible in an isopropanol-filled continuous cloud chamber.

4. Radon counting

Measurement of emanation ionization-induced discharge of an electroscope was the first method to determine its activity. Later on, ionization, scintillation and many other types of detectors were developed for radon counting. Various passive detectors as well as flow-through detectors were developed. Also the direct radon detection techniques, radon progeny detection and even electronic nuclear track detectors were developed.

The direct and precise radon activity determination (typically volume activity in air or water) is, due to radon physical and chemical properties, problematic since the inert gas may escape the sample and the volume activity depends on many factors (e.g., the weather and building ventilation). Thus, particular precautions need to be applied for sample treatment. For example, the well water must be stored in tight bottles in a cold place and measured within few days from the sample collection.

Short-term radon monitoring in air could be easily performed by the measurement of defined amount of air enclosed in a Lucas cell-type detectors, based on ZnS(Ag) scintillation material [46]. Various geometries were developed, and even 1 L cell volume detectors are available with minimal significant volume activities in the range of 2 and 3 mBq/L for airborne radon and ^{222}Rn in water, respectively [47]. Measurement of ^{222}Rn was also performed by its absorption in a plastic scintillators and alpha/beta pulse shape discrimination [48]. These techniques may be used in radon risk determination on a building sites. Soil air is taken from the ground by drilling several exploratory wells up to the depth of 1 m, and air samples are collected with air-tight syringe, transferred to a Lucas cell and counted. Further, the radon gas permeability through the soil and fundaments is also evaluated to determine overall risk [49]. Even though certified methods were applied more detailed analysis is needed in some cases to provide accurate results [50]. For some systems and low-background radon counting, the decayed air or low radon gas is needed to reach low detection limits. Such apparatus for low radon nitrogen was reported [51].

Another type of detectors for longer determination periods (e.g., 2 weeks) are the electret dosimeters [52]. Integral measurement gives an average value of radon volume activity in the air and is less influenced by temporary short-term changes in the monitored place situation.

Long-term determinations are performed with radon nuclear track detectors [53]. These are placed in a monitored area and left for the period of even several years to collect the tracks. These are further etched and detected under microscope and evaluated using CCD camera and PC software [54]. The comparison of various types of alpha-track detectors was evaluated some time ago [55].

For laboratory measurements, semiconductor detectors for alpha and gamma spectroscopy may be used with an advantage of radon, thoron and their progenies discrimination; however, the need of enclosed apparatus is crucial in direct measurements and usually allows only the determination of the progeny. Interesting and a very simple emanation method for determining radium was described, where the radon was adsorbed on a silica gel at the temperature of liquid nitrogen and then transferred at 0°C to a toluene-based liquid scintillator [56].

Other techniques for simultaneous measurement of radon and thoron were studied, for example, using the Timepix electronic nuclear track detector [57]. Electronic radon/thoron detection system was developed, employing the passivated ion-implanted planar silicon (PIPS) detector [58].

A method that allows to distinguish surface radon sources from the deep sources was reported recently [59]. The method for the determination of the relative depth of a radon source is based on the field alpha-spectroscopy of radon (^{222}Rn) and actinon (^{219}Rn) progenies in soil gas. The limitations of the determination were obvious, firstly the 40 s time-period (half-life of actinon) and secondly the geological-structural situation of the studied locality where the high value of the activity ratio of ^{211}Bi and ^{214}Po corresponded to a situation where the short-lived isotope ^{219}Rn was present in the sample in larger amounts than that corresponding to the natural ratio.

Personal monitoring of radiation workers, mainly in uranium industry and other mines in uranium-rich regions, includes a combination of several types of detectors in order to properly estimate the acquired cumulative dose and to discriminate the inhaled radiation burden. Thus they typically contain a thermoluminescent detector for external gamma radiation dose measurement, active flow-through filter unit for the measurement of long-lived progeny aerosols and a nuclear track detector typically equipped with energy-absorbing foils for the radon and thoron alpha decay discrimination [60].

5. Applications and use of radon

The radon element applications include various research, industrial and other fields like its use in spa-based therapies and the exploitation of its natural occurrence in anomalous quantities.

From the early times, radon emanation and the emanation method were applied in the studies of material structure, for example, to determine specific surface, material porosity and their crystalline/amorphous structure [61]. Recently, novel method and an installation for rapid determination of the radon diffusion coefficients in various materials were described recently [62]. Such measurement is important for the development and characterization of radon barriers. Emanation method was also used for the estimation of reactivity of ferric oxides prepared from different sources [63].

To study the radon isotope permeation through the barriers as well as to perform dosimetric studies and other experiments under radon exposure, radon chambers are constructed. A review of different radon chambers with volumes from 0.01 up to 78 m^3 appeared, describing several setups [64]. Fully automated radon chamber was also developed, including the controlled atmosphere (humidity, pressure and radon activity) [65]. The chamber of 1.46 m^3 was made of stainless steel and allowed ^{222}Rn, ^{220}Rn or both to be injected from the bottom pipelines into the chamber in a 100% flow-through mode, 100% recirculate mode or flow-through/recirculate mode.

Another application of radon includes the labelling of surface layers using the active deposits of radon. That is possible, thanks to the nuclear recoil effect and their electrostatic deposition. This method allows to prepare, for example, surface-labelled solid samples and to perform their wear tests [66].

Since the radon gas permeates the underground through the rock cracks and enters into springs and soil air, it may be useful for uranium and other ores prospecting by the emanation detection of radiometric anomalies [67]. Interesting fact of exterrestrial radon occurrence was reported on the lunar surface [68]. Very important geophysical application of radon detection is the measurement of radon release anomalies preceding the earthquakes [69].

Another isotope of radon, the thoron, was used in the separation of ^{212}Pb from ^{228}Th, allowing the construction of a radionuclide generator [70] and its use for the labelling of radioimmunoconjugates for targeted alpha-particle therapy [71]. The production of ^{212}Pb is nowadays performed under good manufacturing practice (GMP), and clinical trials of several tracers are ongoing e.g. [72].

The use of radon (^{222}Rn) in medicine and spas radically decreased as soon as the deterministic effects of ionizing radiation on humans were better understood. On the other hand, under proper regulation, low-activity radon therapy is even nowadays beneficial for the patients suffering from painful inflammatory rheumatic diseases, diseases of musculoskeletal system, diseases of the peripheral nervous system, diabetes and others. In Jáchymov, Czech Republic, the world-famous radium spa is operating four water springs with maximal radon content of 5–20 kBq/L [73]. Despite of insufficient number of clinical trials [74] and some controversies in the radon therapies [75, 76], the long-term experience demonstrably confirms the benefits for the patients and justifies its use, further confirmed by many independent scientific studies [77–80]. The spa medical praxis was verified by decades and is supported also by other epidemiologic studies on the hormesis theory that supports the beneficial effects of low-dose radon exposure [81–83].

Acknowledgements

Author gratefully acknowledges the financial support from the Health Research Agency of the Czech Republic, the Ministry of Interior of the Czech Republic and the Ministry of Education youth and sports of the Czech Republic and the European Communities, under grant agreements no.: NV16-30544A, VI20172020106 and CZ.02.1.01/0.0/0.0/15_003/0000464, respectively.

Author details

Ján Kozempel

Address all correspondence to: jan.kozempel@fjfi.cvut.cz

Department of Nuclear Chemistry, Faculty of Nuclear Sciences and Physical Engineering, Czech Technical University in Prague, Czech Republic

References

[1] Curie M, Curie P, Bémont MG. Sur une nouvelle substance fortement radio-active. Comptes Rendus. 1898;**127**:1215-1217

[2] Schmidt GC. Über die von den Thorverbindungen und einigen anderen Substanzen ausgehende Strahlung. Annalen der Physik und Chemie. 1898;**65**:141-151

[3] Debierne AL. Sur une nouvelle matière radio-active. Comptes Rendus. 1899;**129**:593-595

[4] Curie P, Curie M. Sur la radioactivite provoquee par les rayons de Becquerel. Comptes Rendus. 1899;**129**:714-716

[5] Curie P, Debierne AL. Sur la radio-activité induite provoquée par des sels de radium. Comptes Rendus. 1901;**133**:931-934

[6] Rutherford E, Owens RB. Thorium and uranium radiation. Transactions of the Royal Society of Canada. 1899;**2**:9-12

[7] Rutherford E. A radio-active substance emitted from thorium compounds. Philosophical Magazine. 1900;**49**(296):1-14

[8] Dorn FE. Versuche über sekundärstrahlen und radiumstrahlen. Abhandlungen der Naturforschenden Gesellschaft zu Halle. Stuttgart. 1901;**22**:37-43

[9] Dorn FE. Über die von radioaktiven substanzen ausgesandte emanation. Abhandlungen der Naturforschenden Gesellschaft zu Halle. Stuttgart. 1901;**23**:2-15

[10] Debierne AL. Sur la radio-activité induite provoquée par les sels d'actinium. Comptes Rendus. 1903;**136**:446-449

[11] Rutherford E, Soddy F. The cause and nature of radioactivity. Part II. Philosophical Magazine. 1902;**4**(23):569-585

[12] Rutherford E, Soddy F. Condensation of the radium emanations. Philosophical Magazine. 1903;**5**(29):561-576

[13] Ramsay W, Gray RW. La densité de l'emanation du radium. Comptes Rendus. 1910;**151**: 126-128

[14] Adams EQ. The independent origin of Actinium. Journal of the American Chemical Society. 1920;**42**(11):2205-2208

[15] Gray RW, Ramsay W. Some physical properties of radium emanation. Journal of the Chemical Society, Transactions. 1909;**95**:1073-1085

[16] Chu SYF, Ekström LP, Firestone RB. The Lund/LBNL Nuclear Data Search, Version 2.0. 1999. Available from: http://nucleardata.nuclear.lu.se/toi/ [Accessed: February 21, 2017]

[17] Magill J, Pfennig G, Dreher R, Sóti Z. Karlsruher Nuklidkarte [Chart of the Nuclides]. 9th ed. Eggenstein-Leopoldshafen: Nucleonica GmbH; 2015. ISBN: 978-3-943868-07-4

[18] Vohlídal J, Julák A, Štulík K. Chemické a analytické tabulky. Prague: Grada; 1999 [in Czech]. ISBN: 978-80-7169-855-5

[19] Eichler R, Aksenov NV, Belozerov AV, Bozhikov GA, Chepigin VI, Dmitriev SN, Dressler R, Gäggeler HW, Gorshkov VA, Haenssler F, Itkis MG, Laube A, Lebedev VY, Malyshev ON, Oganessian YT, Petrushkin OV, Piguet D, Rasmussen P, Shishkin SV, Shutov AV, Svirikhin AI, Tereshatov EE, Vostokin GK, Wegrzecki M, Yeremin AV. Chemical characterization of element 112. Nature. 2007;**447**:72-75. DOI: 10.1038/nature05761

[20] Andreyev AN, Antalic S, Huyse M, Van Duppen P, Ackermann D, Bianco L, Cullen DM, Darby IG, Franchoo S, Heinz S, Hessberger FP, Hofmann S, Kojouharov I, Kindler B, Leppanen AP, Lommel B, Mann R, Munzenberg G, Pakarinen J, Page RD, Ressler JJ, Saro S, Streicher B, Sulignano B, Thomson J, Wyss R. α decay of the new isotopes $^{193,194}Rn$. Physical Review C. 2006;**74**:064303. DOI: 10.1103/PhysRevC.74.064303

[21] Kuusiniemi P, Heßberger FP, Ackermann D, Antalic S, Hofmann S, Nishio K, Sulignano B, Kojouharov I, Mann R. Nuclear structure and reactions studies of $^{213g,m}Ra$ and $^{214g,m}Ra$ by α and γ decay. European Physical Journal A. 2006;**30**:551-559. DOI: 10.1140/epja/i2006-10148-y

[22] Crawford JR, Kunz P, Yang H, Schaffer P, Ruth TJ. ^{211}Rn/^{211}At and ^{209}At production with intense mass separated Fr ion beams for preclinical ^{211}At-based α-therapy. Applied Radiation and Isotopes. 2017;**122**:222-228. DOI: 10.1016/j.apradiso.2017.01.035

[23] Meyer G-J, Lambrecht RM. Method for the preparation of radon-211. U.S. Patent No. 4,364,898; December 21, 1982

[24] Bellido AV. New isotopes of emanation and francium: ^{223}Em, ^{224}Em and ^{224}Fr. Journal of Inorganic and Nuclear Chemistry. 1961;**19**(3-4):197-203

[25] Ghiorso A, Meinke WW, Seaborg GT. New low mass isotopes of emanation (element 86). Physical Review. 1949;**76**:1414. DOI: 10.1103/PhysRev.76.1414

[26] Borge MJG, Burke DG, Gabelmann H, Hill P, Jonsson OC, Kaffrell N, Kurcewicz W, Lovhoiden G, Nybo K, Nyman G, Ravn HL, Rogowski J, Thorsteinsen TF. The new neutron-rich isotope ^{228}Rn. Zeitschrift für Physik A: Atomic Nuclei. 1989;**333**:109-110. DOI: 10.1007/BF01290116

[27] Neidherr D, Audi G, Beck D, Blaum K, Böhm C, Breitenfeldt M, Cakirli RB, Casten RF, George S, Herfurth F, Herlert A, Kellerbauer A, Kowalska M, Lunney D, Minaya-Ramirez E, Naimi S, Noah E, Penescu L, Rosenbusch M, Schwarz S, Schweikhard L, Stora T. Discovery of ^{229}Rn and the structure of the heaviest Rn and Ra isotopes from Penning-trap mass measurements. Physical Review Letters. 2009;**102**:112501. DOI: 10.1103/PhysRevLett.102.112501

[28] Chernick CL, Claassen HH, Fields PR, Hyman HH, Malm JG, Manning WM, Matheson MS, Quarterman LA, Schreiner F, Selig HH, Sheft I, Siegel S, Sloth EN, Stein L, Studier MH, Weeks JL, Zirin MH. Fluorine compounds of xenon and radon. Science. 1962;**138**(3537):136-138. DOI: 10.1126/science.138.3537.136

[29] Golovkov NA, Gromova II, Janicki M, Norseyev YV, Sandukovsky VG, Vasaros L. Some radon and astatine compounds produced in a plasma ion source. Radiochemical and Radioanalytical Letters. 1980;**44**(2):67-78

[30] Malli GL. Prediction of the existence of radon carbonyl. International Journal of Quantum Chemistry. 2002;**90**(2):611-615. DOI: 10.1002/qua.963

[31] Tsivion E, Gerber RB. Predicted compounds of radon with acetylene and water. Physical Chemistry Chemical Physics. 2010;**12**(37):11791-11794. DOI: 10.1039/C0CP00875C

[32] Juarez R, Zavala-Oseguera C, Jimenez-Halla JOC, Bickelhaupt FM, Merino G. Radon hydrides: Structure and bonding. Physical Chemistry Chemical Physics. 2011;**13**(6):2222-2227. DOI: 10.1039/C0CP01488E

[33] Rutherford E. Absorption of the radio-active emanations by charcoal. Nature. 1906;**74**:634

[34] Jacobson DR, Khan NS, Collé R, Fitzgerald R, Laureano-Pérez L, Bai Y, Dmochowski IJ. Measurement of radon and xenon binding to a cryptophane molecular host. Proceedings of the National Academy of Sciences of the USA. 2011;**108**(27):10969-10973. DOI: 10.1073/pnas.1105227108

[35] Majer V. Radiochemie. Prague: Jednota českých matematiků a fyziků; 1942 [in Czech]

[36] Parchomenko P. New substances of high emanating power and their application. Part I. Collection of Czechoslovak Chemical Communications. 1938;**10**:54-59. DOI: 10.1135/cccc19380054

[37] Běhounek F. New substances of high emanating power and their application. Part II. Collection of Czechoslovak Chemical Communications. 1938;**10**:60-65. DOI: 10.1135/cccc19380060

[38] Wahl AC, Daniels WR. Emanating power of barium stearate for 3·9-second actinon (^{219}Rn). Journal of Inorganic and Nuclear Chemistry. 1958;**6**(4):278-287. DOI: 10.1016/0022-1902(58)80109-8

[39] Lavi N, Alfassi ZB. Preparation and analysis of ^{226}Ra-^{222}Rn emanation standards for calibrating passive radon detectors. Journal of Radioanalytical and Nuclear Chemistry. 2005;**265**(1):123-126. DOI: 10.1007/s10967-005-0796-4

[40] Eurostandard Ltd. Prague, Czech Republic. Calibration and reference standards. 2011; p. 30. Available from: http://www.eurostandard.cz/Eurostandard-catalog-2011.pdf [Accessed: March 1, 2017]

[41] Hutchinson JMR, Mullen PA, Collé R. The NBS radon-water standard generator. Nuclear Instruments and Methods in Physics Research Section A: Accelerators, Spectrometers, Detectors and Associated Equipment. 1986;**247**(2):385-389. DOI: 10.1016/0168-9002(86)91321-5

[42] Havelka M. Radon-in-water standard. Applied Radiation and Isotopes. 2009;**67**(5):860-862. DOI: 10.1016/j.apradiso.2009.01.047

[43] Bakale G, Rao PS, Mencl J, Adams RB, Evans HH. A radon generator/delivery system. Radiation Research. 1993;**133**(3):277-281. DOI: 10.2307/3578210

[44] Johnson M, Benziger J, Stoia C, Calaprice F, Chen M, Darnton N, Loeser F, Vogelaar RB. A ^{222}Rn source for low-background liquid scintillation detectors. Nuclear Instruments and Methods in Physics Research Section A: Accelerators, Spectrometers, Detectors and Associated Equipment. 1998;**414**(2-3):459-465. DOI: 10.1016/S0168-9002(98)00632-9

[45] Nesmejanov AN, Baranov VI, Zaborenko KB, Rudenko NP, Priselkov JA. Radiochemical Practicum. Prague: SNTL; 1959 [Czech translation]

[46] Lucas HF. Improved low-level alpha-scintillation counter for radon. Review of Scientific Instruments. 1957;**28**(9):680-683. DOI: 10.1063/1.1715975

[47] EMPOS Ltd. Prague, Czech Republic. Lucas cells and scintillators – product detail. Available from: http://www.empos.cz/ionizujici-zareni/mereni-alfa-a-beta-zareni/ds-401-m-ds-404-m/zi_392/50.html [Accessed: March 1, 2017]

[48] Mitev KK. Measurement of ^{222}Rn by absorption in plastic scintillators and alpha/beta pulse shape discrimination. Applied Radiation and Isotopes. 2016;**110**:236-243. DOI: 10.1016/j.apradiso.2016.01.027

[49] National Radiation Protection Institute, Prague, Czech Republic. Radon Programme. 2017; Available from: https://www.suro.cz/en/prirodnioz/rnprogram [Accessed: March 1, 2017]

[50] Goliáš V, Tumurkhuu G, Kohn P, Šálek O, Plášil J, Škoda R, Soumar J. Construction of new houses on a uranium vein outcrop: A case study from the Czech Republic. Nukleonika. 2016;**61**(3):343-349. DOI: 10.1515/nuka-2016-0057

[51] Wójcik M, Zuzel G. Low-^{222}Rn nitrogen gas generator for ultra-low background counting systems. Nuclear Instruments and Methods in Physics Research A. 2005;**539**:427-432. DOI: 10.1016/j.nima.2004.10.023

[52] Kotrappa P, Dua SK, Gupta PC, Mayya YS. Electret—a new tool for measuring concentrations of radon and thoron in air. Health Physics. 1981;**41**(1):35-46. DOI: 10.1097/00004032-198107000-00004

[53] Frank AL, Benton EV. Radon dosimetry using plastic nuclear track detectors. Nuclear Track Detection. 1977;**1**(3):149-179. DOI: 10.1016/0145-224X(77)90011-4

[54] Csordás A, Bátor G, Horváth D, Somlai J, Kovács T. Validation of the scanner based radon track detector evaluation system. Radiation Measurements. 2016;**87**:1-7. DOI: 10.1016/j.radmeas.2016.02.011

[55] Meager WM, Lucas RM, Daum KA, Sensintaffar E, Poppell S, Feldt L, Clarkin M. A performance evaluation study of three types of alpha-track detector radon monitors. Health Physics. 1991;**60**(4):507-515. DOI: 10.1097/00004032-199104000-00004

[56] Darrall KG, Richardson PJ, Tyler JFC. An emanation method for determining radium using liquid scintillation counting. Analyst. 1973;**98**:610-615. DOI: 10.1039/AN9739800610

[57] Janik M, Ploc O, Fiederle M, Procz S, Kavasi N. Optimization of the Timepix chip to measurement of radon, thoron and their progenies. Applied Radiation and Isotopes. 2016;**107**:220-224. DOI: 10.1016/j.apradiso.2015.10.023

[58] Durridge, Inc., Billerica, MA, USA. RAD7 Radon Detector specifications. 2015; Available from: http://www. http://durridge.com/documentation/RAD7%20Specifications.pdf [Accessed: March 1, 2017]

[59] Goliáš V, Pittauerová D, Procházka R, Třískala Z. Field alpha-spectroscopy of radon (^{222}Rn) and actinon (^{219}Rn) progeny in soil gas: Locating a radon source. Journal of Radioanalytical and Nuclear Chemistry. 2005;**266**(3):461-470. DOI: 10.1007/s10967-005-0932-1

[60] IAEA Vienna, Austria. Occupational Radiation Protection in the Mining and Processing of Raw Materials. Safety Guide No. RS-G-1.6. 2004

[61] Hahn O, Graue G. Oberflächenstudien an Gelen des Thoriumoxyds und des Eisenoxyds. Zeitschrift für Physikalische Chemie—Bodenstein Festschrift. 1931;**A**:608-619

[62] Tsapalov A, Gulabyants L, Livshits M, Kovler K. New method and installation for rapid determination of radon diffusion coefficient in various materials. Journal of Environmental Radioactivity. 2014;**130**:7-14. DOI: 10.1016/j.jenvrad.2013.12.010

[63] Balek V. Emanation method for estimating reactivity of ferric oxides prepared from different sources. Journal of Applied Chemistry. 1970;**20**:73-75. DOI: 10.1002/jctb.5010200302

[64] Azimi-Garakani D. A comparison of different radon chambers. Nuclear Instruments and Methods in Physics Research Section B: Beam Interactions with Materials and Atoms. 1992;**71**(1):99-102. DOI: 10.1016/0168-583X(92)95347-T

[65] Leung JKC, Jia D, Tso M-YW. A fully automated radon exposure chamber. Nuclear Instruments and Methods in Physics Research Section A: Accelerators, Spectrometers, Detectors and Associated Equipment. 1994;**350**(3):566-571. DOI: 10.1016/0168-9002(94)91259-9

[66] Jech Č, Kubašta J, Gosman A. Electrostatic deposition of thoron decay products used for labelling of surface layers. Journal of Radioanalytical and Nuclear Chemistry. 1998;**230**(1-2):281-283. DOI: 10.1007/BF02387480

[67] ÚSVTRS (Central Administration of Research and Exploitation of Radioactive Materials). Catalogue of Anomalies. 1955 [in Czech/Russian]

[68] Lawson SL, Feldman WC, Lawrence DJ, Moore KR, Elphic RC, Belian RD, Maurice S. Recent outgassing from the lunar surface: The Lunar Prospector Alpha Particle Spectrometer. Journal of Geophysical Research. 2005;**110**:E09009. DOI: 10.1029/2005JE002433

[69] Chalov PI, Tuzova TV, Alekhina VM. Pulsed nature of helium and radon content anomalies in some water sources of North-Tien Shan seismic zone in period preceding Kochkor earthquake of 1974. Doklady Akademii Nauk SSSR. 1976;**231**:1331-1334

[70] Hasfjell S. A ^{212}Pb generator based on a ^{228}Th source. Applied Radiation and Isotopes. 2001;**55**:433-439. DOI: 10.1016/S0969-8043(00)00372-9

[71] Wrasidlo W, Mysels KJ. Generation of lead-212 and radioimmunoconjugates for use in cancer therapy and the 'emanating power' of stearates. Colloids and Surfaces A: Physicochemical and Engineering Aspects. 2001;**186**:123-128. DOI: 10.1016/S0927-7757(00)00768-8

[72] Meredith RF; Torgue JJ; Rozgaja TA; Banaga EP; Bunch PW; Alvarez RD; Straughn JMJr; Dobelbower M C; Lowy A M. Safety and outcome measures of first-in-human intraperitoneal α radioimmunotherapy with 212Pb-TCMC-Trastuzumab. American Journal of Clinical Oncology. 2016; DOI: 10.1097/COC.0000000000000353

[73] Léčebné lázně Jáchymov. Jáchymov, Czech Republic. Léčivé zdroje (in Czech). Available from: http://www.laznejachymov.cz/lecive-zdroje/ [Accessed: March 1, 2017]

[74] Franke A, Reiner L, Pratzel HG, Franke T, Resch KL. Long-term efficacy of radon spa therapy in rheumatoid arthritis—A randomized, sham-controlled study and follow-up. Rheumatology. 2000;**39**:894-902

[75] Swiss Federal Commission for Radioprotection and Radioactivity Surveillance. Opinion Concerning Radon Therapy for Ankylosing Spondylitis. Approved on December 5, 2014. Available from: https://www.bag.admin.ch/dam/bag/en/dokumente/str/srr/ksr-stellungnahme-radontherapie.pdf [Accessed: May 1, 2017]

[76] Zdrojewicz Z, Strzelczyk J. Radon treatment controversy. Dose-Response. 2006;**4**(2):106-118. DOI: 10.2203/dose-response.05-025.Zdrojewicz

[77] Moder A, Hufnagl C, Jakab M, Hitzl W, Ritter M. Radon-therapy in ankylosing spondylitis reduces auto-antibody titers. Open Journal of Molecular and Integrative Physiology. 2011;**1**:52-54. DOI: 10.4236/ojmip.2011.13008

[78] Kanzaki N, Kataoka T, Etani R, Yamaoka K. Recent studies on anti-inflammatory effects of radon inhalation in mice. Inflammation & Cell Signaling. 2015;**2**:e601. DOI: 10.14800/ics.601

[79] Pauthner M, Janko C, Baum W, Schorn C, Dobias H, Moder A, Schett G, Herrmann M. Radon therapy ameliorates disease progression and prolongs survival in TNF α TG mice. Annals of the Rheumatic Diseases. 2012;**71**(Suppl 1):A1-A93. DOI: 10.1136/annrheumdis-2011-201233.1

[80] Moder A, Dobias H, Ritter M. Effects of low-dose radon therapy applied under hyperthermic conditions (RnHT) on inflammatory and non-inflammatory degenerative disease conditions. In: Huilgol N, editor. Hyperthermia. Rijeka: Intech; 2013. DOI: 10.5772/51401. ISBN: 978-953-51-1129-0

[81] Fornalski KW, Dobrzyńsky L. The cancer mortality in high natural radiation areas in Poland. Dose-Response. 2012;**10**:541-561. DOI: 10.2203/dose-response.11-035.Fornalski

[82] Thompson RE. Epidemiological evidence for possible radiation hormesis from radon exposure: A case-control study conducted in Worcester, MA. Dose-Response. 2011;**9**:59-75. DOI: 10.2203/dose-response.10-026.Thompson

[83] Scott BR. Residential radon appears to prevent lung cancer. Dose-Response. 2011;**9**:444-464. DOI: 10.2203/dose-response.11-027.Scott

Chapter 3

Radon Monitoring in the Environment

Abd Elmoniem Ahmed Elzain

Additional information is available at the end of the chapter

http://dx.doi.org/10.5772/intechopen.69902

Abstract

Radon is a natural radioactive gas used to estimate the radioactive hazard in the environment. Radon (222Rn), which is one of the daughters of uranium (238U), represents the most essential isotope, with a half-life of 3.825 days. The associated health risks due to inhalation and ingestion of radon and its progeny when present in enhanced levels in an indoor environment like a human dwelling have been documented. In this chapter, we have discussed the sources and techniques besides the methods used for measuring radon gas in the environment including soil, water, building materials, etc., which are well documented. A wide range of techniques for the detection and quantification of radon has been developed over the last few years. There is no single technique that can meet all the requirements of the different types of the radon measurements. Finally, we have mentioned the most essential information effecting the radon monitoring in the environment and the methods of measuring and controlling the concentration values throughout the environment; we have also mentioned the effect of radon on the inhabitants through the estimation of the effective dose rates and lung cancer risk due to radon gas when its values exceed the action level values.

Keywords: indoor radon, radioactivity dose rate, sources of radon, relative risk of lung cancer, passive and active techniques

1. Introduction

Radon is a radioactive noble gas that does not chemically react with other elements. However, it can change the physical properties of the surrounding medium. Radon (^{222}Rn), which is one of the daughters of uranium (^{238}U), represents the most essential isotope, with a half-life of 3.825 days. Its half-life allows it to migrate long enough to travel long distances and accumulate into the indoor environment. Radon (^{222}Rn) and its short-lived decay products

(^{218}Po, ^{214}Pb, ^{214}Bi and ^{214}Po) in dwellings are recognized as the main sources of public exposure from the natural radioactivity, contributing to nearly 50% of the global mean effective dose to the public [1]. The interest in studying radon behaviour is mainly due to the fact that it can accumulate indoors and in case of entering into the body can have serious damage to the human respiratory and gastrointestinal systems. The human respiratory tract due to radiation, radon and its daughter nucleus after entering the body is exposed to the most damage, producing an increased risk to the population. Therefore, after smoking, the second factor of lung cancer is radon [2]. The associated health risks due to inhalation and ingestion of radon and its progeny when present in enhanced levels in an indoor environment, water and soils surrounding a human dwelling have been well documented [3–7]. The radon concentration is usually depending on many parameters: including the radium content in the soil, meteorological parameters and the radon emanating from various soils and rocks [8, 9].

The radon concentration in the ground depends on the radium content of the soil and the emanation power of soils and rocks [10]. Among many factors affecting radon exhalation, one of the most important is radium content of the bedrock or soil. However, radon exposure shows an extreme variation from location to location and depends primarily on the exhalation rate of radon from the soil. It was recorded that radon exhalation rate studies are important for understanding the relative contribution of the material to the total radon concentration found inside the dwellings [11–13]. Radon and its daughters are emitted from building materials; one source of radon in houses is the building materials. The radon exhalation rate studies for building material samples are important for understanding the relative contribution of the material to the total radon concentration found inside the dwellings. It was shown that the radon exhalation rate decreases with the building age [14–16].

Radon contained in water is to some extent transferred into room air as a result of agitation or heating [17, 18]. UNSCEAR 2000 reported that radon concentration as a rule is much lower in surface water than in groundwater; radon concentrations in groundwater are expected to be 40 Bq L^{-1}; the mean dose from radon in water from inhalation was 0.025 mSv y^{-1} and 0.002 mSv when swallowed [1]. According to the WHO, the mean of the radon concentrations in tap water from surface waters equals 0.4 Bq L^{-1} and in well water is 20 Bq L^{-1} [19]. The release of radon from water to air depends upon circumstance in which the water is used, as the degassed fraction increases considerably with temperature. A recent report of large radon concentration in drinking water has added a new element of deep concern to the problem of environmental health hazards [20–22]. Measurements of radon concentration in water have mostly been undertaken in regions where high levels were suspected. It was tentatively estimated in the UNSCEAR 1982 Report (Annex D, Paragraph 163) that between 1 and 10% of the world's population consumes water containing radon concentrations of the order of 100 Bq L^{-1} or higher, drawn from relatively deep wells. For the remainder, who consume water from aquifers of surface sources, the weighted world average concentration is probably less than 1 kBq m^{-3} [23]. The International Commission for Radiological Protection has suggested that areas where 1% or more of the building have indoor radon concentration 10 times higher than that of national average should be considered as "radon prone" areas. It is also recommended that radon concentration value ranges 500–1500 and 200–600 Bq m^{-3} for work places and dwellings, respectively; those concentrations do not pose a significant risk for workers [2]. Therefore, it is desirable not only to measure the radon but also to find out the sources of radon especially in the houses [4]. Other sources such as natural gas also contribute towards radon activity in dwellings, which depends upon their origin and the rate of consumption.

2. Sources of radon

2.1. Indoor sources

The studies of indoor radon levels have been conducted with long- and short-term methods in order to assess the indoor radon problems [24–26]. Indoor radon is influenced by many sources, such as soil, building materials, water and natural gas. A correct calibration procedure is paramount for good accuracy of results. Hence, precalibrated techniques either passive or active were used to study indoor radon-222 concentrations. These techniques were used in monitoring radon concentration through the entire environment [27, 28]. For such techniques as the solid state nuclear track detectors (SSNTDs), track density could be determined and then converted into activity concentration C_{Rn} (Bq m^{-3}) using the following equation [29]:

$$C_{Rn} = \frac{\rho_{Rn}}{K_{Rn}T} \quad (1)$$

where ρ_{Rn} is the track density (tracks per cm^2), K_{Rn} is the calibration constant, which must be previously determined in tracks cm^{-2} h^{-1} per (Bq m^{-3}) and t is the exposure time.

2.2. Outdoor sources

The quantity of radon in the earth is very small and amounts to about 4×10^{-7}% by weight. The main source of radon outdoors is the radium in the earth crust; at least 80% of the radon emitted into the atmosphere comes from the top few metres of the ground [30, 31]. The concentration of uranium and radium in the ground varies with types of rocks and minerals. The world average concentration for soils is about 24 Bq kg^{-1} for uranium-238. The concentration of radium in rocks and soil is often (but not always) the same as that of uranium.

2.3. Soil sources

2.3.1. Emanation

The emanation of radon atoms from a material is the process by which the radon atoms escape from a quantity of the material. The mechanisms of radon release from rock, soil and other materials are not very well understood and are probably not the same. It was reported that the emanation rate of radon is influenced by the condition of soil and its porosity, moisture content, temperature and atmospheric pressure [23, 32]. If the moisture content is very low, then the radon release is decreased by the effect of re-adsorption of the radon atoms on surfaces in the pores. On the other hand, if the moisture content increases slightly, the radon release increases up to a certain moisture content, above which the release of radon decreased again owing to the decreasing diffusion rate in water-filled pores [33]. It was found that the emanation of radon from soil depends not only on the ^{238}U and ^{226}Ra concentration but also on the nature of the host mineralogy and the permeability of the host rock and the soil. It was found that the effect of paint, in general, is to reduce the radon emanation from the brick surfaces, which increases the radon concentration inside the brick. The emanation of radon increases from the unpainted area of the brick due to the increase of radon concentration inside it. It was also found that the radon emanation increases from brick covered with gypsum and plaster; therefore, the internal finish can increase the radon concentration inside houses [34].

2.3.2. *Diffusion*

Diffusion is identified as the principal method by which radon is transferred into and out of the basement modules and appears to be relatively independent of insulating materials and vapour retarder. It was found that the variability of radon and correlations with differential pressure gradients may be related to air current in block walls and soil that interrupted radon diffusion inward. Once radon has entered the air or water surrounding the emanating radium-bearing particle, it is transported by diffusion, earth mechanical and convective flow, percolating of rainwater and flow of groundwater. It was shown that there is insufficient evidence to accept the pressure-driven mechanism as the dominant mechanism of radon infiltration in homes, and thermo-diffusion gas flow in clay and concrete can greatly exceed the pressure-driven flow [35].

2.3.3. *Flux from the soil*

The radon emanation and exhalation rates in soil vary from place to another due to differences in radium concentration and soil parameters, such as moisture content, porosity, permeability and grain size. The moisture content of the soil is defined as the mass of water in the soil expressed as a fraction of the dry weight of the solids forming the soil matrix [36, 37]:

$$\%\ \text{moisture content} = \frac{\text{mass of soil water}}{\text{mass of dry soil solids}} \times 100 \tag{2}$$

$$\%\ \text{moisture content} = \frac{W_1 - W_2}{W_2} \times 100 \tag{3}$$

where W_1 and W_2 are the masses of the wet and dry samples.

The porosity p of any porous material is a measurement of the total pore space that can be occupied by water in that material and can be expressed as % given by [36]

$$\text{porosity} = \frac{V_w - V_d}{V_w} \times 100 \tag{4}$$

where V_d and V_w are the volume of dry and wet samples:

$$V_d = \frac{\text{Weight of the dry samples}}{\text{Density of the soil grain}} \tag{5}$$

The diffusion equation of radon through soil, with an assumption that the medium is without source, is given by [36]

$$\frac{D_e}{p}\frac{d^2C}{dx^2} - \lambda C = 0 \tag{6}$$

where C is the concentration of the gas, λ is the decay constant of the gas in s^{-1}, D_e is the effective diffusion coefficient of the gas in the porous medium in $cm^2\ s^{-1}$ and p is the porosity of the medium.

The solution of Eq. (6) is

$$C(x) = C_\circ \exp\left(\left(-\sqrt{\frac{\lambda p}{D_e}}\right)x\right) \tag{7}$$

where $C(x)$ is the concentration of radon at any time t at a distance x, C_o is the concentration of the radon in the source and $D_e = D \times p$ where D is the diffusion coefficient. Relations between the radon concentration in soil and the indoor radon concentration are rather complicated. Nevertheless, the data on radon concentration in the soil is the starting point for the assessment of the expected radon on the constructed building. It was shown that the most relevant soil parameters on the radon flux at the top of the crack are, in this case, effective diffusion coefficient, soil gas permeability and deep soil radon concentration [38, 39]. Radon concentration from an infinite column of soil as a function of distance z into soil is given by [40]

$$C(z) = C_\infty(1 - \exp(-z/d)) \tag{8}$$

where $d = \sqrt{D/\lambda}$, λ is the decay constant, D is the diffusion constant and C_∞ is the concentration at infinite depth.

2.4. Building material sources

Radon emitted from building materials. The natural building materials with higher radium content are of large concern. The type of aggregate material, water and natural gas contribute significantly in an indoor concentration levels. This may depend on the consumption rate and the origin of the source. Measurements of radon exhalation rates could be done by both active and passive techniques [7].

2.4.1. Calculation of radon concentration and radon exhalation rates

For the purposes of calculating ^{222}Rn concentration and radon exhalation rate, when the study was conducted by using the passive diffusion dosimeter, the surface exhalation rate of radon E_x ($Bq\ m^{-2}\ h^{-1}$) is determined by Eq. (9), as [12, 13]

$$E_x = \frac{\lambda VC}{A\left[T + \lambda^{-1}(\exp(-\lambda T) - 1)\right]} \quad (9)$$

The mass exhalation rate of radon E_M (mBq kg^{-1} h^{-1}) is determined by using the following (Eq. (10)) [7, 13]

$$E_M = \frac{\lambda VC}{M\left[T + \lambda^{-1}(\exp(-\lambda T) - 1)\right]} \quad (10)$$

where C is the calculated integrated radon exposure as measured by SSNTDs in (Bq m^{-3} h), T is the exposed period of the detector (hours), V is volume (m^3), λ is the decay constant of radon (h^{-1}), A is the surface area (m^2) and M is the sample's mass (kg).

2.5. Water sources

Radon and radium in water are exposed by either ingestion from the daily consumption of water or by inhalation through the daily routine use of water. They constitute a health risk for inhabitants [5, 41]. Due to clathrate behaviour, the solubility of radon decreases with temperature, but temperature dependence is much stronger for heavier gases. Radon can be transported by groundwater to far larger distances than by diffusion process in a short time [42].

2.5.1. *Calculation of radon activity density from water samples*

The radon activity density in water C_w was calculated using the formula [43, 44]

$$C_W = \frac{\lambda Cht}{L} \quad (11)$$

where λ is the decay constant of radon-222, h is the distance from the surface of water in the sample can to the detector, t is the exposure time of the sample and L is the depth of the sample.

The time elapsed for the sample collection and analysis is corrected using the following equation:

$$C_W = C_0 e^{-\lambda t} \quad (12)$$

where C_W is the measured concentration, C_0 initial concentration (to be calculate) after the decay correction and t is the time elapsed since collection (days), $\lambda = 0.181$, $t_{1/2} = 3.83$ days.

2.6. Dispersion in air

The dispersion of radon in air is influenced by the vertical temperature gradient, the direction and strength of the wind and the air turbulence [45]. The dispersion of radon daughters is also influenced by precipitation and washout ratios. The radon progeny concentrations and washout ratios are inversely correlated with low precipitation intensities [46]. It was found that the correlation between radon concentration and wind speed shows a broad inverse correlation.

The explanation of this correlation is the dispersion of radon in a larger volume due to vertical mixing under stable atmospheric conditions and the dural variations. It was also found that the effect of turbulence drops the radon concentration using a mixing fan [47].

2.7. Mines and mine tailings

Sources of radon of local interest include the tailings from uranium mining, milling and geothermal power stations. The radon exhalation rate from tailings depends on the radon content of the tailings, the emanation factor and the land reclamation [48]. The required uranium is mined in the form of uranium oxides; the richer ores contain some 1–4 kg uranium per 1000 kg of ore [49]. Mining of all kinds affects the environment negatively. In the area of mining, the specific concerns are the occupational radiation exposure of miners and population living in the vicinity of the mining area. Miners are exposed to airborne radon and its short-lived decay products, alpha emitters and gamma dose rate. Radon emanation from a uranium mine is an important pathway for public radiation exposure [50]. The tailings consist of residues with still enough uranium to produce radon gas escaping into the surroundings; in fact 85% of the radioactive materials of the ore are still in tailings [49].

2.8. Natural gas and crude oil

In nature, the sources of oil and gas are sedimentary rocks, namely, sands, sandstone, grills, limestone and dolomites. The sedimentary and igneous rocks also contain trace amounts of uranium-238 in varying quantities which is the source of radon [23]. With the oil exploration, ^{238}U may also be extracted, which is present in crude oil and natural gas.

Facilities producing natural gas and condensate are mainly affected by radon and one of its decay product polonium-210, with half-life of 138.4 days. The boiling point of radon is approximately 62°C, and it will tend to follow gases of similar boiling point (ethane and propane) through the separation circuits. Radon will progress through the facility in the gas stream, by its decay products, which tend to plate-out on pipelines and vessel surfaces. The 138.4-day half-life of polonium-210 allowed this radionuclide to build to sometimes very high levels, which can only be detected by using alpha particle monitoring instruments. Radium and its decay products can also present in silt materials deposited in the separator, dehydration and drying units and condensate recovery vessels. Some facilities use a polonium treatment unit during condensate recovery to remove polonium isotopes, and the polonium concentrations in these units can become very high over a period of years of operation [51].

In oil production both radon gas and its direct precursor radium-226 can be a problem. Radon in the stream will be separated by the gas scrubbers. The problem is similar to the gas plant, in that polonium-210 will plate-out on an available surfaces. However, radium-226 can also be present as a dissolved solid and will act with another chemical in the liquid phase, to form insoluble compounds, which are deposited as a scale in pipe work and vessels. This scale can be detected by gamma radiation monitors externally but may be misleading with alpha radiation, because of the deposition of polonium-210 from radon. Scale can be presented in all areas of the process plant and will gradually increase with time. In some cases, the radioactivity of scales and silts is high enough to warrant that they be deposited of low specific activity radioactive wastes [51].

2.9. Dose estimation and lung cancer risk

Regarding radiation dose to the public, due to waterborne radon, it is believed that waterborne radon may cause higher risk than all other contaminants in water. Therefore, radon in water is a source of radiation dose to stomach and lungs. The annual effective doses for ingestion and inhalation were calculated according to parameters introduced by UNSCEAR report [1].

The annual effective dose due to inhalation corresponding to the concentration of 1 Bq L^{-1} in tap water is 2.5 μSv y^{-1}. In the UNSCEAR, 2000, report, the annual effective dose rate E_{Eff} could be calculated using the following (Eq. (1)):

$$E_{Eff}\left(\text{mSv y}^{-1}\right) = C_{Rn}\left(\text{Bq m}^{-3}\right) \times 8760\left(\text{h y}^{-1}\right) \times 0.4 \times 0.8 \times 9.0 \times 10^{-6}\left(\text{Bq m}^{-3}\,\text{h}^{-1}\right) \quad (13)$$

The annual effective dose rate was also related to the average radon concentration C_{Rn} by the following expression [1, 7]:

$$E_{Eff}\left(\text{WLM y}^{-1}\right) = \frac{8760 \times n \times F \times C_{Rn}}{170 \times 3700} \quad (14)$$

where C_{Rn} is the radon concentration in Bq m^{-3}, n is the fraction of time spent indoors, F is the equilibrium factor, 8760 is the hours per year and 170 is the hours per working month. For purposes of the effective dose equivalent estimation, a conversion factor of 6.3 mSv WLM^{-1} should be used [1].

The annual effective dose to an individual consumer due to intake of radon from drinking water is evaluated using the relationship [3]:

$$E_W\left(\text{Sv y}^{-1}\right) = C_W \times C_{RW} \times D_{CW} \quad (15)$$

where E_w is measured in (Sv y^{-1}), C_w is the concentration of radon in water (Bq L^{-1}), C_{Rw} is the annual consumption of water (L y^{-1}) and D_{cw} is the dose conversion factor for radon (Sv Bq^{-1}). Following exposure of radon from consumed water, annual effective doses (μSv y^{-1}) and effective doses per litre (nSv L^{-1}) could be calculated [3].

The relative risk of lung cancer (RRLC) due to indoor exposure to radon should be calculated by using the following formula [29, 52]:

$$\text{RRLC} = \exp(0.00087352\, C_{Rn}) \quad (16)$$

3. Materials and methods

3.1. Introduction

A wide range of techniques for the detection and quantification of radon and its daughters have been developed over the last few years. There is no a single technique that can meet all the requirements of the different types of the radon measurements. The choice of the most appropriate one depends on the particular information needed, the type of radon surveys and the cost of the apparatus. For example, a large number of long-term integrated measurements are needed for the measurements of the population exposure to radon, while continuous radon monitoring is required to study the dependence of the soil radon on the environmental and geophysical parameters.

3.2. Short-term radon measurements

Short-term measurements (from a few days to 1 month) are particularly for screening surveys to identify houses with high radon concentrations and to investigate the geographical variations. When such short-term measurements are employed for the assessment of radon exposure in dwellings, particular care is needed to define a protocol of exposure. To this end, activated charcoals, electret ion chambers (EICs) and solid state nuclear track detectors (SSNTDs) are the most attractive techniques.

Different attempts have been made to improve the response of activated charcoal devices, especially with the use of diffusion barrier to decrease the effect of humidity. With the electret ion chamber (EIC) devices, the most important limitation is their sensitivity to gamma radiation, which, however, can be easily corrected with any means including an additional radon proof electret ion chamber. SSNTDs are not efficiently sensitive for short-term measurements. This limitation is due essentially to the small area normally counted. Another alpha track detector, which is not sensitive to the plate-out, is polycarbonate and can be conveniently used as a bare detector [53].

3.3. Long-term radon measurement

The most representative measurements of the true exposure of radon gas in any environment are the long-term (up to 1 year) integrating types of measurements.

The most commonly used detectors are, respectively, SSNTDs and electrets. These detectors are used as bare detectors or in combination with closed radon samplers. The most established geometry for a closed radon sampler consists in a chamber with a porous filter such as fibre glass, a non-wetting cloth or a micropore paper filter [54] and one or more SSNTDs. These filters, which are used to stop external radon daughters, do not discriminate thoron gas and water vapour.

3.4. Active and passive measurement

The measurements of radon/radon daughters by passive techniques are based on the detection of their radioactive decay.

There are different ways to define active and passive types of radon measurement systems. The classification, which seems to avoid any possible confusion, is reported in the following:

3.4.1. Active instrumentation and/or methods

The sampling of radon/radon daughters is made by forced sampling through the use of power supply (pumps).

3.4.2. Passive instrumentation and/or methods

The sampling of radon/radon daughters is based on the natural diffusion without any power supply. Another important classification can be introduced according to the type of radiation detector used:

- Detectors with real-time response (typically scintillators and semiconductors).
- Detectors with no real-time response (e.g. track-etched detectors, activated charcoal and electrets).

In practice a radon/radon daughter monitor may be formed by any possible combination of active or passive sampling systems and real-time or no real-time detectors.

3.5. Methods of measurements of radon gas concentration

Radon measurement techniques are performed for different applications. The radon concentration in air is highly dependent on several factors, mainly the air ventilation. The various detectors one can put in a bore hole for radon concentration measurements are essentially solid state nuclear track detectors, daughter collectors such as alpha card, electret detectors and thermoluminescent phosphors and solid state electronic detectors such as photodiode and gas absorber like activated charcoal [55].

3.5.1. Direct measurements

Passive sampling can be combined with real-time detectors for direct (not integrated) or continuous measurements. For example, in the continuous passive radon detectors, radon diffuses through a filter in a ZnS (Ag)-coated cell optically coupled with a photomultiplier tube. Passive sampling is followed by electrostatic collection in combination with scintillation discs and photomultiplier tube or silicon surface barrier detectors [56].

Within the silicon technology several new devices, such as random access memories (RAMs), photodiodes or charged coupled devices (CCDs), although developed for different applications, can be conveniently used for alpha particle detection and for radon monitoring. The low power requirements, the low cost and the high reliability of these devices make them uniquely interesting for passive radon monitoring [57].

3.5.2. Passive absorption measurements

Radon gas can be collected by adsorption on activated charcoal and then measured using gamma detectors or liquid scintillation technique passive detectors as thermoluminescent detector (TLD) can also be used in combination with activated charcoal. Short-term integrating monitoring using activated charcoal has been extensively employed in large surveys [58].

3.5.3. Passive diffusion, electrostatic collection and passive detectors

In this system radon is sampled by diffusion in a chamber, and its daughters are collected by an electrostatic field on aluminized Mylar film facing the scintillation disc of ZnS (Ag). The alpha particle that induced the scintillations is registered by photographic films, a similar passive system referred to as Environmental Gamma Ray and Radon Detector (EGARD) has been reported by Maiello and Harley [59], in which system the radon daughters are collected through an electret-induced electric field and registered by a thermoluminescent detector (TLD).

3.5.4. Passive sampling and passive detectors

Totally passive devices can successfully be obtained for radon-only measurements in which radon diffuses through a filter or a membrane in a detector housing, and the radiations from the radon daughter produced are directly registered by passive type of detectors.

Different types of passive detectors can be used such as thermoluminescent materials, electret devices and solid state nuclear track detectors (SSNTDs). With respect to large-scale surveys, SSNTDs have the most favourable characteristics for radon measurements and their applications. The most often used track detectors available for the registration of alpha particle are essentially cellulose nitrate (typically the red-dyed LR-115), bisphenol-a poly carbonate (Makrofol or Lexan) and allyl diglycol carbonate (CR-39). CR-39 nuclear track detector method is based on letting air diffuses in a holder where alpha particles are detected by means of a solid state nuclear tack detector [60].

3.6. Physical bases of radon gas measuring techniques

3.6.1. Solid state nuclear track detectors

Basically, solid state nuclear track detectors (SSNTDs) work as follows:

The passage of heavily ionizing particles, such as alphas, through most insulating solids creates narrow paths of intense damage on an atomic scale. These damage trails may be enlarged until they can be seen under an optical microscope, by chemical treatment that rapidly and preferentially removes the damaged material. It removes less rapidly the surrounding undamaged matrix in such a manner as to enlarge the etched tracks that mark and characterize the site of the original damaged region [61]. SSNTDs are sensitive to alpha particles but totally insensitive to beta and gamma rays. They do not require energy to be operated and unaffected by humidity, low temperature, moderate heating and light. After irradiation, SSNTDs are chemically processed, and the determination of the number of particles that have impinged the detector can be performed by various means. The most common is the counting of the tracks under an optical microscope, but many other techniques have been devised [61].

3.6.2. Alpha card

According to its producer alpha nuclear, the alpha card system is a passive radon detecting method, which provides a sensitive measure of radon in gaseous state. The card is deposited in the ground, suspended in an inverted cup. It is left in place from 12 h up to several days. After entering the volume of detection, the radon decays away, and its daughters make an active deposit on a thin membrane. The alpha card is then recovered and read. The reading device contains two silicon detectors, sensitive to alpha particles only, facing one another and between which the card is inserted. In this manner a good counting efficiency is achieved. The reader is small, portable equipment that can be used in the field.

3.6.3. *Electret detectors*

One of the most recent techniques is the electret radon monitor. The electret dosimeter offers several advantages: its ability to store information over a relatively long period, its independence of the humidity in its environment and its easiness of reading. But its response curve does not cover efficiently the very low or very high doses, and also it is sensitive to gamma radiation background, which in some cases induces significant error or even precludes their uses.

3.6.4. *Activated charcoal*

The adsorption of radon could be measured in an activated charcoal using a plastic can. The dosimeter should be replaced in the location where measurement is intended, lid open and left in place for 4–12 days. It is then retrieved and the lid closed and brought to the lab where the gamma activity is determined.

3.6.5. *Thermoluminescent detectors (TLDs)*

Based upon the principle of light emission after heating, several thermoluminescent dosimeters (TLDs) have been constructed and improved. The idea of detecting radon by using these dosimeters is depending on the ability of recording the alpha activity of the radon daughters. After a suitable exposition, the TLD is recovered and "read" in a TLD reader [62].

3.6.6. *Solid state electronic detectors*

Electronic detectors for radon measurements in the soil are not widely used because they usually require energy and also they are rather expensive and fragile. One may mention the alpha-metre manufactured by Alpha Nuclear Company; the active part of the equipment is a Si (Li) detector associated with a counter unit. The counting is directly displayed on the instrument.

Using photodiode simple equipment that can be operated on pen-type batteries for a year has been developed. The detector does not require a polarization voltage, and the simplifying electronics is a very low consumption setup. Data are stored in memory cards of a microcomputer. The main advantage of electronic equipment of this type is that they allow dynamic studies of low term variations on the radon concentration [63].

3.6.7. *Pylon chamber*

Filtered air is pumped into an electrostatic chamber in which the negative electrode is made of a thin conductive Mylar foil placed on a ZnS scintillator for alpha particle detection. Three hours after the end of sampling, the alpha particles emitted by radon and radon daughters are counted [63].

3.6.8. *Solid state alpha spectrometry*

A known volume of air is forced through a filter. The radon daughter's concentration in the air sample is measured through alpha spectrometry on the filter using a solid state detector [63].

3.6.9. *Active carbon detectors*

Air diffuses in a canister where radon gas is adsorbed by active carbon. Sampling lasts 2–4 days; later the gamma rays emitted by the radon progeny captured into the active carbon are counted by means of scintillation detector (ROLS, 4300E USA).

3.6.10. *Lucas cells*

Glass or transparent plastic holders (cells) internally coated with ZnS are used. Just before sampling the cell is evacuated. Then, the air to be sampled is filtered and let into the cell. After at least 3 h delay, the cell is placed in optical contact with a photomultiplier and alpha particles emitted by radon, and its daughters are counted. The sensitivity of the system is about 4 Bq m^{-3} [55].

3.6.11. *Markov method*

This method is based on two successive counting of alpha particles from a glass fibre filter through which a known air volume has been previously aspirated. This technique supplies the measure of the potential alpha energy concentration [63].

3.7. Sampling methods for the measurement of radon in air

Methods and instrumentation may be classified on the basis of the type of monitoring as grab sampling and continuous and time-integrating monitoring.

3.7.1. *Grab sampling methods*

3.7.1.1. *Radon sampling and counting scintillation flask*

The scintillation cell technique is one of the most widely used in various applications. The device consists essentially of a container with a transparent bottom coated internally with silver-activated zinc sulphide phosphor. The bottom is coupled to a photomultiplier tube and counting device. Cells sizes ranges from 0.10 to 2.0 L. The scintillation flasks can be filled either by evacuation or by airflow using a pump.

The lower level of detection (LLD) depends on several parameters, such as the size of flasks, type of material and counting intervals. For typical devices a sensitivity of 3.7 Bq m^{-3} is reported [63].

3.7.1.2. Ionization chambers

Ionization chambers are very complex laboratory instrument essentially used in laboratory as highly accurate measurement systems, which are directly traceable to reference standards. An ionization chamber consists of two conductors with a glass-filled space and an electric field applied between them. Pulses from collector are fed into the preamplifier and from here to a linear amplifier and scalar. The collector anode is connected to a device sensitive to voltage changes. The instrument can be operated either by alpha pulse counting or by ionization current measurement. Special techniques are used for filling the counter with filtered air to remove atmospheric aerosols including radon daughter products. Flow-through ionization chambers can be used for both laboratory applications, and as field instruments, sealed ionization chambers are essentially laboratory instruments. Samples are collected in the field and returned for laboratory assay where they are transferred ionization chambers for counting. To test performance and efficiency of the instruments, including the filling apparatus, a sample of radon is drawn from a standard source of ^{226}Ra. The theoretical sensitivity is of the order of 10–4 A Bq^{-1} at equilibrium Rn/Rn_D [63].

3.7.1.3. Two-filter method

The device consists of a cylindrical tube, which is fitted to both ends with filter holders and a pump to draw air. The first filter allows entering only radon and other atmospheric gases. The exit filter collects on the inside radon daughters formed within the chamber.

The activity of the outlet filter, which is proportional to the radon concentration in air, is measured by alpha counting devices: ZnS (Ag) scintillation detectors or solid state barrier detectors. The two-filter technique can be used for short-term sampling in workplaces or for continuous outdoor monitoring of very low levels of radon concentrations [63].

3.7.2. Continuous sampling instruments for radon air monitoring

Continuous monitoring of radon concentration is used both for research and radioprotection purposes. The main applications concern diagnostic of radon sources for remedial actions, the measurement of variations of ambient radon concentration and soil gas radon levels. Two different methods of sampling are generally used.

3.7.2.1. Diffusion sampling continuous radon monitor

The diffusion radon monitor is a device that does not require power supply for sampling, but makes use of electronic circuitry for counting alpha particles. Radon gas enters into the volume of the instrument by molecular diffusion through a filtering device.

Radon progeny resulting from the decay of radon within the sensitive volume is counted by scintillation or a surface barrier detector. To improve sensitivity Rn_D atoms can be focused of the detector surface with an electrostatic collector. The efficiency of the electrostatic collection depends on particle motilities, which are reduced by the aggregation of water molecules around the ions in the humid atmosphere. Beds of silica gel are used to make these devices independent of the moisture content of the air [63].

3.7.2.2. Flow-through sampling continuous radon monitor

Air is drawn continuously through double stopcock scintillation flask by an air pump. The total counts registered in a given interval are a function of the radon concentration, the activity of daughters collected during the interval and the activity of RnD collected during previous intervals. This effect is of particular importance in the presence of rapid variation of radon concentration. With changing radon concentration, the radon progeny build-up inside the scintillation flasks should be taken into consideration during the calibration. This effect requires a correction if instruments are calibrated with steady-state radon concentrations. Calibration factors could overestimate or underestimate real radon concentrations. A special device based on a solid state barrier detector was developed for monitoring fast changes of radon concentrations [64].

3.8. Time-integrating radon monitors

The time-integrating monitors measure the average radon concentration during the exposure period. Various devices have been developed based on the diffusion of radon into a sensitive volume through a filtering device and the measurement of radiations emitted by radon and its progeny using as active counting devices either scintillation ZnS (Ag) or silicon surface barrier detector [65].

3.9. Radon monitoring in soil

Plastic track detectors, viz. cellulose nitrate films (LR-115 type 2) manufactured by Kodak-Pathe, France, and CR-39 sheets (Pershore Moulding Co., UK), are very sensitive for recording alpha particle tracks produced by radon. Plastic track detectors are capable of recording average value of radon isotopes over relatively longer periods (weeks to months) but are insensitive to transient variations that last only several hours or less.

3.9.1. Track-etch method

Track-etch technique is quite appropriate for radon detection in the soil gas because of its negligible background of spurious signals, low cost, ruggedness and nature as an integrating measurement. The detector assembly rests in the vertical position near the bottom of auger hole and radon decay alpha particles impinge on the detector films leaving their radiation damage trials. LR-115 films could be collected after a week or month, as the case may be, and etched in a constant temperature bath using 2.5 N NaOH solutions for 2 h at 60°C. Alpha tracks are revealed as circular or conical spots, which become etched-through holes after prolonged etching. The track spots or holes would be counted using a binocular microscope under a suitable magnification. The measured track density is assumed to be proportional to the average radon content in soil gas [66].

3.9.2. Radon emanometry

Radon emanometer (type RMS-10) is used to measure the instantaneous radon concentration in the soil gas. The apparatus consists of an alpha counting scintillation assembly with inverted bell-shaped alpha detector, a hand-operated rubber pump and a soil-gas probe. The probe is metallic tube about 4 cm diameter, with perforations at the lower end and a rubber capping at the top to seal it pneumatically in an auger hole. It has inlet and outlet tubes for the circulation of the soil gas. The hand-operated rubber pump is used to circulate the soil gas into the scintillation chamber. The alpha particles emanating from the radon impact the ZnS (Ag) scintillators creating an energy impulse in the form of light quanta, which are recorded by scintillation assembly [67].

3.9.3. Silicon-diffused junction detector

Alpha metre-400 is designed to measure near surface radon gas fluctuations. It consists of silicon-diffused junction for detection of alpha particles and gives sufficient counts over 24 h exposures in most of the soils. The detector unit is placed inside a covered auger hole about 60 cm in depth. The detector is separated from the soil surface at the bottom of hole by a 6.4 cm gap, which shields the detector from the impact of the direct alpha particles emanated by radon isotopes and their alpha emitting daughters [68].

3.10. Radon monitoring in groundwater

Radon emanometry technique has been used for groundwater radon measurements. The radon content is determined by measuring radon-generated alpha activity in discrete water samples collected once a day from the monitored wells and natural springs.

3.10.1. Radon measurements in groundwater using RAD7

Radon measurement of groundwater samples could be used by radon-in-air monitor RAD-7. The monitor is used to determine radon-in-air activity concentrations by detecting the alpha-decaying radon daughters (Po-218 and Po-214) using a passivated implanted planar silicon detector (PIPS). At the end of the run, the RAD7 prints out a summary, showing the average radon reading [21].

4. Conclusions

From our study we can conclude that, we have mentioned the most important information about radon throughout the environment, the health importance of prompt radon daughters, the radon problem, the risk of lung cancer and the radon measuring techniques. Then we discussed the main sources of radon in the environment (indoor, soil, building materials, water, etc.) and the techniques used for measuring radon in the environment including long-term and short-term radon measurements and active and passive measurements; then we summarizes the methods of measurements of radon gas concentration, physical basis of radon gas measuring techniques, sampling methods for the measurement of radon in the air, time integrated radon monitors, radon monitoring in soil, radon monitoring in groundwater and passive diffusion radon dosimeters (CR-39). This chapter presents the essential and important information in a concise manner in order to clarify the importance of radon gas in the environment.

Author details

Abd Elmoniem Ahmed Elzain[1,2]*

*Address all correspondence to: moniemelzain1@yahoo.com

1 Department of Physics, Kassala University, Sudan

2 Department of Physics, College of Science and Arts in Oqlat Alsqoor, Qassim University, Saudi Arabia

References

[1] UNSCEAR. United Nations Scientific Committee on the Effects of Atomic Radiations, Sources and Effects of Ionizing Radiation, Vol. I Annex A: Dose Assessment Methodologies, United Nations, New York; 2000

[2] ICRP, International Commission on Radiological Protection Against Radon-222 at Home and at Work; ICRP Report 65, Pergamon: Oxford; 1993

[3] UNSCEAR. United Nations Scientific Committee on the Effects of Atomic Radiations, and effects of ionizing radiation. Report to the General Assembly, with Scientific Annexes, United Nations, New York; 1993

[4] Elzain AEA. A study of indoor radon levels and radon effective dose in dwellings of some cities of Gezira State in Sudan. Nuclear Technology and Radiation Protection. 2014; **29**(4):307-312

[5] Elzain AEA. A study on the radon concentrations in drinking water in Kassala State (Eastern Sudan) and the associated health effects. World Applied Sciences Journal. 2014;**31**(3):367-375

[6] Elzain AEA. Estimation of soil gas radon concentration and the effective dose rate by using SSNTDs. International Journal of Science and Research Publications (IJSRP). 2015;**5**(2): 3887-3891

[7] Elzain AEA. Radon exhalation rates from some building materials used in Sudan. Indoor and Built Environment. 2015;**24**(6):852-860

[8] Idriss H, Salih I, Abdulaziz AS, Abdelgalil MY, Salih SA, Hasan AM, ELtahir MA, Ahamed MMO. Study of radon in soil gas, trace elements and climatic parameters around South Kordofan state, Sudan. Environmental Earth Sciences. 2014;**72**(2):335-339

[9] Nazaroff WW, Nero, Jr. AV. Radon and its Decay Products in Indoor Air. New York: John Wiley and Sons; 1988

[10] Akerblom G, Wilson C. Radon-Geological Aspects of an Environmental Problem. Report No. 30. Uppsala: Geological Survey of Sweden; 1982

[11] Elzain AEA. Determination of radium concentration and radon exhalation rate in soil samples using CR-39. Advances in Applied Science Research. 2015;**6**(2):92-102

[12] Elzain AEA, Mohammed, YSh, Mohamed KhSh, Mohamed-Ali AM. A study of radium concentration and radon exhalation rate in soil samples from Kassala Town, Sudan using SSNTDs. American Journal of Physics and Applications. 2016;**4**(4):84-89

[13] Elzain AEA, Mohammed YSh, Mohammed KhSh, Sumaia SM. Radium and radon exhalation studies in some soil samples from Singa and Rabak Towns, Sudan using CR-39. International Journal of Science and Research (IJSR). 2014;**3**(11):632-637

[14] Abu-Jarad F. Application of nuclear track etch detector for radon related measurements. International Journal of Radiation Applications and Instrumentation Part D Nuclear Tracks and Radiation Measurements 1988;**15**(1):525-534

[15] Khan AJ, Parasad R, Tyagi RK. Measurements of radon exhalation rate from some building materials. Nuclear Tracks and Radiation Measurements. 1992;**20**(4):609-610

[16] Yu KN, Chan TF, Young ECM. The variation of radon exhalation rates from building surfaces of different ages. Health Physics. 1995;**68**(5):716-718

[17] Karpińska M, Jacek K, Zenon M, Andrzej S. Radon in drinking water in the Białystok region of Poland, NUKLEONIKA. 2010;**55**(2):177-180

[18] Mowlavi AA, Shahbahrami A, Binesh A. Dose evaluation and measurement of radon concentration in some drinking water sources of the Ramsar region in Iran. Isotopes in Environmental and Health Studies. 2009;**45**(3):269-272

[19] WHO. Guidelines for Drinking Water Quality. 3rd ed. Geneva; 2008

[20] Abdul Kader Arafin SK. Radon concentration in shallow and deep tube-well water samples using SSNTD method. Online Journal of Physical and Environmental Science Research. 2012;**1**(1):1-4

[21] Badhan K, Mehra R and Sonkawade RG. Measurement of radon concentration in ground water using RAD7 and assessment of average annual dose in the environs of NITJ, Punjab, India. Indian Journal of Pure and Applied Physics. 2010;**48**(7):508-511

[22] Nikolov J, Todorovic N, Forkapic S, Bikit I, Mrdja D. Radon in drinking water in Novi Sad. Proceedings of World Academy of Science, Engineering and Technology. 2011;**76**: 307-310

[23] UNSCEAR. Exposure to Radon and Thoron and Their Decay Products. Report to the General Assembly, Annex D. United Nations, New York, USA; 1982

[24] Alshahri F and Alqahtani M. Radon concentrations and effective radium contents in local and imported phosphate fertilizers, Saudi Arabia. Arabian Journal for Science and Engineering. 2015;**40**:2095-2101

[25] Mukhtar OM, Elzain AEA. Radon monitoring at Khartoum using the charcoal technique. Abhath al-Yarmouk. Basic Sciences and Engineering. 2006;**15**(2):225-235

[26] Espinosa G, Gammage RB. Long-term radon survey in Mexico City and other regions of the country. Radiation Measurements. 1997;**28**(1-6):663-665

[27] Alter HW, Price PB. Radon detection using track registration material. US Patent, 3, 665, 194 Terradex Corp; 1972

[28] Elzain AEA. Indoor Radon-222 concentrations in some Cities in Kassala State, Eastern Sudan. In: Proceedings of the 2011 International AASRT Symposium, Orlando, Florida, USA. Vol. 2; 2011. pp. 71-84

[29] Elzain AEA. Measurements of indoor radon levels and dose estimation and lung cancer risk determination for workers in health centres of some towns in the Sudan. American Journal of Modern Physics. 2016;**5**(4):51-57

[30] Nazaroff WW. Radon transport from soil to air. Review of Geophysics. 1992;**30**(2):137-160

[31] NCRP. Exposure from Uranium Series with Emphasis on Radon and its Daughters. Bethesda, MD: National Council on Radiation Protection and Measurements, Report No. 77; 1984

[32] UNSCEAR. Sources, Effects and Risks of Ionizing Radiation. United Nations Report, New York, U.S.A: United Nations Publications E. 94 IX. 11; 1988

[33] ICRP. International Commission on Radiological Protection, Recommendations of the International Commission on Radiological Protection, ICRP Publication 26, Annals of the ICRP, 1(3), Pergamon Press, Oxford; 1977. pp. 11-13

[34] Abu-Jarad F, Fremlin JH. Effect of internal wall covers on radon emanation inside houses. Health Physics. 1983;**44**(3):243-248

[35] Minkin L. Is diffusion, thermodiffusion, or advection a primary mechanism of indoor radon entry? Radiation Protection Dosimetry. 2002;**102**(2):153-161

[36] Shweikani R, Giaddui TG and Durrani S. The effect of soil parameters on the radon concentration values in the environment. Radiation Measurements. 1995;**25**(1-4):581-584

[37] Gundersen LCS. The effect of rock type, grain size, permeability and moisture content on radon on soil gas. Journal of Radioanalytical and Nuclear Chemistry. 1992;**161**:325-337

[38] Mazur D, Janik M, Loskiewicz J, Olko P and Swakon J. Measurements of radon concentration in soil gas by CR-39 detectors. Radiation Measurements. 1999;**31**:295-300

[39] Albarracín D, Font LI, Amgarou K, Domingo C, Frenández F and Baixeras C. Effect of soil parameters on radon entry into a building by a means of the TRANSRAD numerical model. Radiation Protection Dosimetry. 2002;**102**(4):359-364

[40] Flischer RL and Magro-Campero A. Mapping of integrating radon emanation for detection of long-distance migration of gases within the earth: Techniques and principals. Journal of Geophysical Research. 1978;**83**(B7):3539-3549

[41] Cross FT, Harely NH and Homann W. Health effects and risks from radon-222 in drinking water. Health Physics. 1985;**48**:649-670

[42] Menin A, Drigo A, Capra E and Verri G. Radon monitoring in dwellings and in water in localized areas of Friuli-Venezia Giulia (Italy). In: Furlan G and Tommasino L, editors. Proceedings of the 2nd Workshop on Radon Monitoring in Radioprotection, Environmental and/ or Earth Sciences, ICTP, Trieste, Italy, 1991, Singapore: World Scientific; 1991. pp. 79-87

[43] Somogyi G, Hafez A, Hunyadi I and Toth–Szilagyi M. Measurement of exhalation and diffusion parameters of radon in solids by plastic track detectors. Nuclear Tracks And Radiation Measurements. 1986;**12**(1-6):701-704

[44] Elzain AEA. Measurement of radon-222 concentration levels in water samples in Sudan. Advances in Applied Science Research. 2014;**5**(2):229-234

[45] Akber RA, Pfitzner J and Johnston A. Wind direction correlated measurements of radon and radon progeny in atmosphere: A method for radon source identification. Radon and Radon Progeny in Australia. Symposium, Canberra; 1994; 18 Feb. pp. 55-63

[46] Paatero J. Wet deposition of radon-222 progeny in Northern Finland measured with an automatic precipitation gamma analyser. Radiation Protection Dosimetry. 2000;**87**(4): 273-280

[47] Abu-Jarad F and Sextro RG. Reduction of radon progeny concentration in ordinary room due to the mixing fan. Radiation Protection Dosimetry. 1988;**24**(1/4):507-511

[48] IAEA. Measurements and Calculations of Radon Releases from Uranium Mill Tailings. Technical Report Series No. 333. Vienna: International Atomic Energy Agency; 1992

[49] Egbert B and Rienk Van G. Environmental Physics. England: John Wiley & Sons Ltd.; 1995

[50] IAEA. The Application of the Principals for Limiting Release of Radioactive Effluents in the Case of Mining and Milling of Radioactive Ores. Safety Series No. 90. Vienna: International Atomic Energy Agency; 1989

[51] Chandler WP. Radon gas in oil and natural gas facilities. Radon and Radon Progeny in Australia. Symposium, Canberra; 1994; 18 Feb. pp. 71-75

[52] Elzain AEA, Sam AK, Mukhtar OM, Abbasher MA. A survey of indoor radon-222 concentration measurements in Kassala Town, Gezira. Journal of Engineering & Applied Sciences. 2008;**3**(2):72-100

[53] Rannou A, Jeanmarie L, Tymen G, Mouden A, Naour E, Parmentier N, Renouard H. Use of cellulose nitrate as radon and radon daughter detectors for indoor radon measurements. Nuclear Tracks and Radiation Measurements. 1986;**12**:747-750

[54] Fleischer RL. Radon in the environment—opportunity and hazards. Nuclear Tracks and Radiation Measurements. 1988;**14**(4):421- 435

[55] Bam BB, Bassigani A, Finazzi P, Giacomlli G, Lembo L, Maltoni G, Partizii L, Predieri F, Sartorio C, Serra P, Steinberg R, Togo V. Radon measurement in the Gran Sasso

underground laboratory. In: Furlan G, Tommasino L, editors. Proceedings of the 2nd Workshop on Radon Monitoring in Radioprotection, Environmental and/or Earth Sciences; 25 Nov.-6 Dec. 1991; ICTP, Trieste, Italy. Singapore: World Scientific; 1991. pp. 492-498

[56] Negro VC, Watnick S. FUNGI—A radon measuring instrument with fast response. IEEE Transactions on Nuclear Science NS. 1978;**25**:757

[57] Shott JV. Ionizing particle detection by charged coupled devices (CCDs). In: Furlan G, Tommasino L, editors. Proceedings of the 2nd Workshop on Radon Monitoring in Radioprotection, Environmental and/or Earth Sciences; 25 Nov–6 Dec. 1991; ICTP, Trieste, Italy. Singapore: World Scientific; 1991. pp. 257-268

[58] Prichard HM, Marien K. Desorption of radon from activated charcoal into a liquid scintillator. Analytical Chemistry. 1983;**55**:155

[59] Maiello ML, Harley NH. EGARD: An environmental gamma-ray and radon detector. Health Physics. 1987;**53**:301

[60] Beozzo M, et al. Measurement of indoor radon levels in Bologna metropolitan area. In: Proceedings of the 15th International Conference on; Particle Tracks in Solids, Marburg, Germany; 1990. p. 297

[61] Fleisher RL, Price PB, Walker RM. Nuclear Tracks in Solids-Principles and Application. Berkeley: University of California Press; 1975

[62] Yamamoto K, Fujii Y, Icotooka Y, Katayama T., Highly stable silicon pin photodiode, Nuclear Instruments and Methods in Physics A. 1987;**253**:542-547

[63] Bochicho F, Risica S. Active radon and radon daughter monitors. In: Tommasino L, Furlan G, Khan AH, Monnin M. editors, Proceedings of the Workshop on Radon Monitoring in Radioprotection, Environmental Radioactivity and Earth Sciences; April 3-14; ICTP, Trieste, Italy. Singapore: World Scientific; 1989. p. 110

[64] Wantick S, Latner N, Graveson RT. A radon-222 monitor using alpha spectroscopy. Health Physics. 1986;**50**:645

[65] Wrenn ME, Spitz H, Cohen N. Design of a continuous digital output environmental radon monitor. IEEE Transactions on Nuclear Science. NS. 1975;**22**:645

[66] Elzain AEA. Assessment of indoor radon doses received by the students and staff in schools in some towns in Sudan. International Journal of Science and Research (IJSR). 2015;**4**(1):2566-2571

[67] Ghosh PC, Bhalla NS. A close circuit technique for radon measurement in water and soil, with some of its applications. In: Proc. All India Symp. On Radioactivity and Metrology of Radionuclides, AEET, Bombay; 1966. p. 14

[68] Warren RK. Recent advances in uranium exploration with electronic alpha cups. Geophysics. 1977;**42**:982-989

Chapter 4

Levels of Radon and Granite Building Materials

Akbar Abbasi

Additional information is available at the end of the chapter

http://dx.doi.org/10.5772/intechopen.69987

Abstract

Natural radioactivities in granite building materials that are commonly used in the world have been surveyed two type methods: a passive measurement method and an active measurement method. In the passive measurement method, after sample preparation, the concentration of ^{226}Ra is measured in the samples. The measuring procedure carried out by using a gamma-ray spectrometry system with a high purity germanium detector. Also, in the active measurement method, an alpha GURD model PQ 2000 radon meter is used directly. In this method, a standard room is considered ((4.0 m × 5.0 m area × 2.8 height) that ground and walls have been covered with granite stones, we calculated radon concentration and radon exhalation rate. Moreover, the results of the exhalation rates measured by passive and active methods are compared and the results are the same, within errors better than 22%.

Keywords: building materials, natural radioactivity, granite, radon, exhalation

1. Introduction

Radioactive radon-222 (^{222}Rn) gas has a half-life of 3.8 days, which emanates from rocks and soils and tends to concentrate in enclosed spaces like underground mines or houses. Radon is a major contributor to the ionizing radiation dose received by the general population [1].

Most terrestrial materials contain ^{238}U and radon gas emitters from these materials, since ^{222}Rn is a decay product of Ra-226, which in turn is taken from the longer lived prior U-238. Concentrations of ^{238}U and ^{226}Ra in some terrestrial materials such as alum shale and black shale are high. Certain granites are typical of radioactivity-bearing natural materials, but it is always feasible to find natural radioactivity-rich bedrocks of different kinds such as construction materials. Construction materials that used as building materials are sources of indoor airborne radioactivity and external radiation from the decay series of uranium in buildings [2].

Short-lived radioelement outcomes from radon are the most important contributors to human exposure to ionizing radiation from natural sources. This contribution represents 50% of the total annular human dose. The parameters such as relative humidity, airflow, type of building materials, indoor-outdoor temperatures and ventilation rate as well as geological formations affect indoor radon concentrations. It is important to estimate the impress and contribution of the different building materials that can deed as Rn sources or Rn absorber inside of buildings for residence and work.

According some reports, the granite stones have a higher radon exhalation rate than other terrestrial materials on average [3–6].

2. Granite building materials

The most abundant of plutonic rock in mountain belts and continental shield areas were made by granite stones. Types of batholith stones that may in dwell thousands of (km^2) are usually intimately associated with granodiorite, quartz, gabbro and diorite. They are scratch resistant and extremely durable; their hardness lends themselves for the stone to be mechanically polished to a high gloss finish. They mainly contain of large grains of potassium, sodium feldspars and quartz. Granite is a plutonic rock in which quartz makes up between 10 and 50% of the felsic components and alkali feldspar accounts for 65–90% of the total feldspar content. Applying this definition requires the mineral identification and quantification abilities of a competent geologist. Other components of granites stones are hornblende and mica. A typical granite rock, in chemical view, is composed of 5% soda ($NaHCO_3$), 5% potassium oxide (KO_2), 12% aluminium (Al), 75% silica (SiO_2), as well as lime (CaO), iron (Fe), magnesia ($Mg(OH)_2$) and titanium (TiO_2) in minor quantities.

In general, it was widely believed that granite stones were formed mostly from magmatic differentiation of basaltic magma. The evidence of this exegesis was considered to indicate a metamorphic origin [7].

Figure 1 illustrates the generalized mineral composition of igneous rocks. Granites and rhyolites (compositionally equivalent to granite but of a fine grain size) are composed mainly of orthoclase feldspar, quartz, plagioclase feldspar, mica and amphibole.

2.1. Radon-222 in granite building materials

In nature, two types of radioactivity sources including technologically enhanced naturally occurring radioactive materials (TENORMs) and naturally occurring radioactive materials (NORMs) consist of materials, usually industrial wastes or by-products enriched with radioactive elements found in the environment, such as U, Th and K and any of their decay products, such as Ra and Rn. All natural products with terrestrial origin, especially granite stones, sand and minerals, include trace amounts of some NORM radioactive elements that can produce measurable amounts of radiation and sometimes Rn-222. This includes all concrete products, clay bricks, most non-plastic plates and dishes, coal and the fly ash produced in coal-fired power plants, natural gas (contains Rn), phosphate fertilizers used in your arable land (all contain K and small amounts of U and Th) and the vegetables grown using those fertilizers. All glasses made by using silica (even wine glasses, eye glasses, windows, mirrors, etc.) and granite stone too. **Figure 2** shows the percentage of radon source in residential place. As seen in figure, only 2.5% of radon gas is released from building materials, whereas granite building material has highest contribution to radon exhalation rate among other materials.

A fraction of the radon activity produced by decay of ^{226}Ra in building materials enters buildings by diffusion. The area exhalation rate can be expressed as

$$R = \lambda_{Rn}\rho_{Build}C_{Build,\,Ra}\,F_r L_{Rn} tagh\left(\frac{L_h}{L_{Rn}}\right) \tag{1}$$

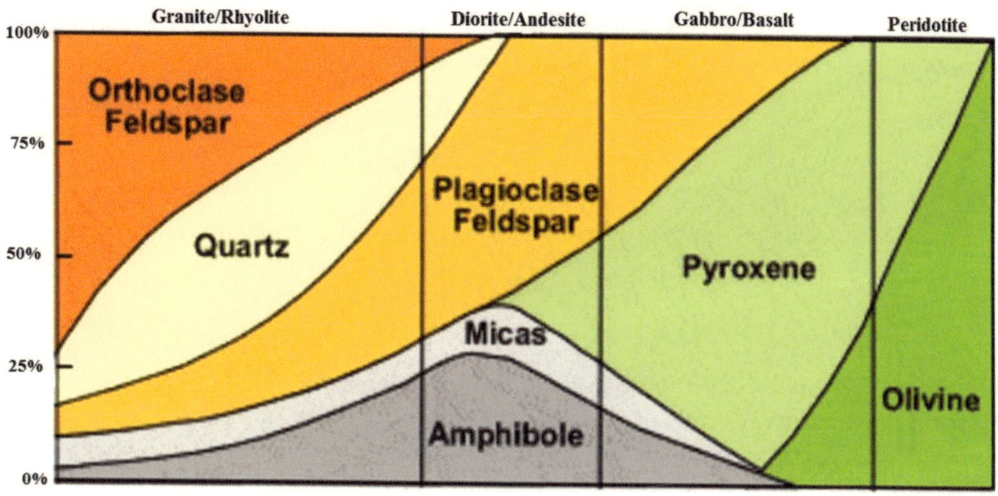

Figure 1. Granite composition diagram.

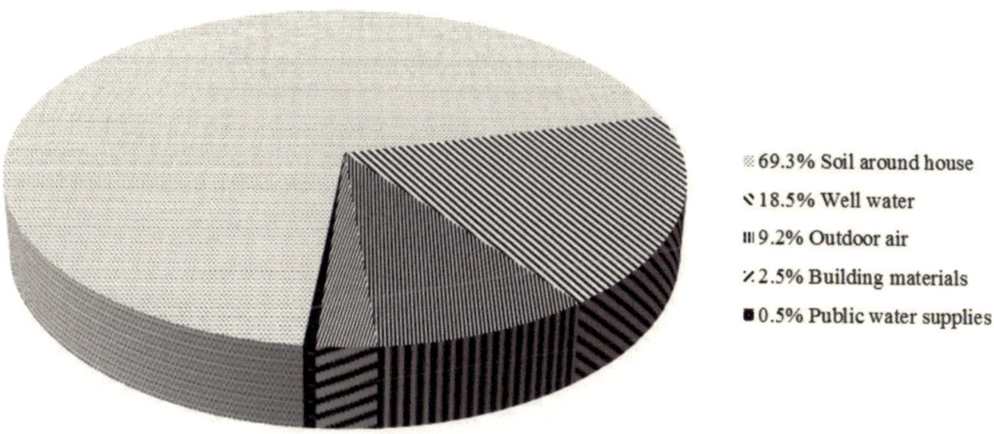

Figure 2. Radon source contribution in residential places.

which is an equation very similar to that related to soil, the only difference being the introduction of a hyperbolic term to account for the fact that diffusion takes place in a medium of finite thickness.

In this equation R is the area exhalation rate (Bq m^{-2} s^{-1}), F is the emanating power, ρ_{build} is the density of the building material (kg m^{-3}), $C_{Build, Ra}$ is the activity mass concentration of ^{226}Ra in the building material (Bq kg^{-1}), L_{Rn} is the diffusion length (m) and L_h is the half thickness of a slab of building material (m).

The mass activity exhalation rate expressed in Bq kg^{-1} s^{-1} and defined as

$$R_m = \lambda_{Rn} C_{Build, Ra} F_r \tag{2}$$

is the quantity usually determined in laboratory measurements. The area exhalation rate from a wall or a floor made of building material of half-thickness L_h, diffusion length L_{Rn} and density ρ_{Build} can then be expressed as

$$R = R_m \rho_{Build} L_{Rn} tagh\left(\frac{L_h}{L_{Rn}}\right) \tag{3}$$

The rate of entry of radon resulting from exhalation from building materials may be expressed as

$$U_{Bm} = \frac{N}{V}(2R_c S_F + R_b S_w) \tag{4}$$

where N is the number of seconds per hour, V is the volume of the reference house (250 m^3), S_F is the surface area of its floor or ceiling (100 m^{-2}), S_w is the surface area of its external walls (100 m^2) and R_c and R_b are the area exhalation rates from concrete and brick, respectively.

3. Measurement of radon-222 in granite building materials

Radon-222 concentration can be measured by calculation of exhalation rate from material surface. We offer two different methods to calculate the radon-222 concentration in building materials: passive method and active method.

3.1. Passive method

3.1.1. Theoretical approaches

In theoretical view, we consider a defined model. In this model, we assume that the release from materials brought into the room is negligible and radon gas is homogeneously mixed with the room air. Then, the concentration in a room can be found by solving the following equation:

$$\frac{\partial C_i(t)}{\partial t} = E_X \frac{S}{V} + C_0 \lambda_v - C_i(\lambda_{Rn} + \lambda_v) \tag{5}$$

where $C_i(t)$ is the ^{222}Rn radioactivity concentration in the home at time t in Bq m^{-3}, E_X is the ^{222}Rn exhalation rate per Bq (m^{-3} h^{-1}), S is the surface area that Rn gas is exhaling (m^2), V is the bulk of home (m^{-3}), λ_{Rn} is the ^{222}Rn decay constant and λ_v is the rate of air exchange at time t in h^{-1}.

This value range is among 0.1 h^{-1} and 3 h^{-1}for residence as air exchange rate value of 0.5 h^{-1} is suggested for residential mechanical ventilation systems (MVSs) according by UNSCEAR report [8]. The value C_0 is the outside ^{222}Rn concentration in the world average value (WAV) 10 Bq m^{-3} in the outside air [9]. At the steady state $\left(\frac{\partial C_i(t)}{\partial t} = 0\right)$, the ^{222}Rn concentration in the room is given by Anjos et al. [10]

$$C_{i=} \frac{E_x S/V + C_0 \lambda_v}{(\lambda_{Rn} + \lambda_v)} \tag{6}$$

From the measured values of ^{226}Ra concentration, the radon exhalation rate per unit area can be calculated by

$$E_x = \frac{1}{2} A_{Ra} \lambda \rho \eta d \tag{7}$$

where ρ is the material density (kg m^{-3}), d is the wall thickness (m) and η is the emanation coefficient. The emanation coefficient value was to be reported between 6.04 and 2.54% (average: 4.29%) for granite stones [11].

The presence of water in building materials alters the transport condition; therefore, the above equation is valid only for dry conditions [12].

3.1.2. *External γ-ray radiation dose rate*

The activity concentration and indoor γ-ray radiation dose rate are calculating for a rectangular source shape with uniform density. The external γ-ray dose rate is calculating in the middle of a standard room with dimension (5.0 m × 4.0 m × 2.8 m) by summing the separately calculated γ-ray dose rates caused by walls and floor. The specific indoor dose rates depended on a large wall thickness and density of materials. It does not depend on the position in the room and dimensions of the room [13].

The density of granites stones is calculating by the experimental method in laboratory. This value approximately is 2580 kg m^{-3}, on average. The granite in markets is usually 3.0 cm thick and (30.0 cm × 50.0 cm) dimension. Also, the dose rate conversion factor is calculated by calculations based on the point kernel integration method for floor covered with 3.0 cm thick granite [14]. The free-in-air absorbed dose value in the middle of the room can be expressed as [10]

$$D(nGy.\,h^{-1}) = K_K A_K + K_{Ra} A_{Ra} + K_{Th} A_{Th} \tag{8}$$

where D is absorbed dose in the center of the room, A_{K}, A_{Ra} and A_{Th} are the activity concentration (Bq kg^{-1}) of ^{40}K, ^{226}Ra and ^{232}Th, respectively. Coefficients of K_{K}, K_{Ra} and K_{Th} are their dose conversion factors or specific dose rates in nGy h^{-1} per Bq kg^{-1}.

3.1.3. Experimental procedure

3.1.3.1. Sample preparation

The collected samples are pulverized, sieved through 0.2 mm mesh, sealed in standard 1000-ml Marinelli beakers, dry-weighed and stored for 4 weeks before counting in order to allow the reaching of equilibrium between ^{226}Ra and ^{222}Rn and its decay products. It is assumed that the radionuclides in equilibrium, i.e. the activity of each daughter, were equal to the initial isotope of the series.

3.1.3.2. Sample counting

The γ-ray spectra of the prepared samples are measured using a γ-detector. This detector is a typical high-resolution γ spectrometer based on a coaxial P-type shielded high purity germanium (HPGe) detector. Efficiency of this detector is 80%with a relative photopeak and energy resolution is 1.80 keV full-width at half-maximum (FWHM) for the 1332 keV γ-ray spectra of ^{60}Co, coupled to a high count-rate multi-task 16k multi-channel analysis (MCA) card. Also, the Gamma-2000 commercial software is used for data analysis. For sample counting, each sample is put into the lead shielding container of the HPGe detector and measured for a collect time of 24 hours. Prior to the samples measurement, the environmental background γ-radiation at the counting place site is determined with a blank Marinelli beaker under identical measurement conditions. It is later subtracted from the measured γ-ray spectra for each sample. The ^{232}Th, ^{238}U, ^{226}Ra and ^{40}K activity concentrations are calculated for each of the measured samples together with their corresponding total uncertainties. ^{232}Th activity concentration are measured by taking the mean activity of photopeaks of the decay nuclides ^{228}Ac (968, 911 and 338 keV) and ^{212}Pb (238 keV). On the other hand, ^{226}Ra activity concentrations are calculated from the activity of its short-lived daughter's radionuclide ^{214}Pb at 351 and 295 keV and ^{214}Bi at 609.6 keV. Activity concentrations of ^{40}K are determined directly from its gamma emission at 1460 keV [17].

To compare results of uncertainty, standard materials such as soil-375 and RGK-1 (K_2SO_4), RGU-1 (U-ore) and RGTh-1 (Th-ore) are used, and all obtained results in accordance are completed. The activity levels of ^{226}Ra and ^{232}Th in soil-375 are 424, 20 and 20.5 Bq kg^{-1}, respectively. Also, activity levels of ^{40}K in RGK-1, ^{238}U in RGU-1, ^{232}Th in RGTh-1 are 14,000, 4940, 3250 Bq kg^{-1}, respectively.

For lower limit of detection (LLD), 96% confidence is calculated using a relation of LLD = 4.66 $(F_c)^{1/2}$, where F_c is the Compton background in the region of the selected gamma-ray spectrum.

Figure 3 shows a type of high purity germanium detector (HPGe) set-up and accessories parts.

Figure 3. A high purity germanium detector (HPGe) set-up used in the passive method.

3.2. Method of active measurement

To measure radon-222 gas in the active measurement method, an alpha GURD model PQ-2000 detector is used. This device works based on the passing of radon-222 gas from a filter to an ionization chamber. Rn-222 surveys are carried out in a special cubic chamber (70 × 50 × 60 cm) with different changeable walls. The walls have different covering materials on the internal surfaces. One set of floor is covered with the most common granite stones. Each sample is put into a special cubic chamber in floor and half walls and measuring for an accumulating time of

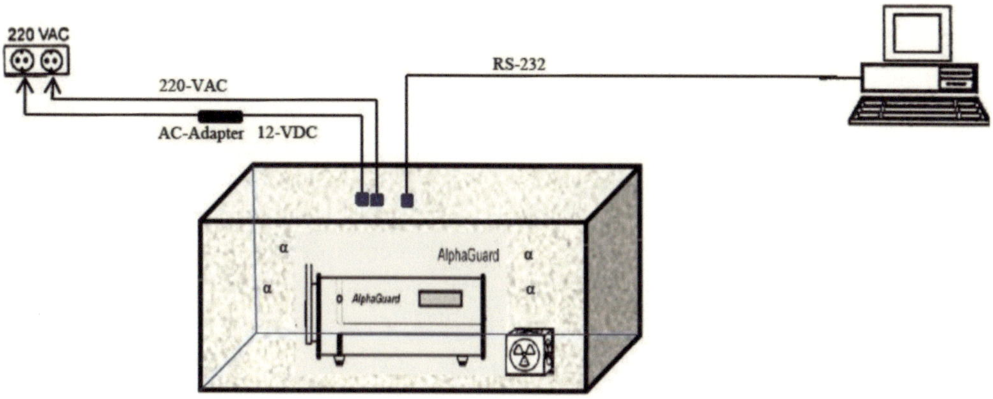

Figure 4. Schematic diagram showing the radon exhalation measurements of granite samples by the active setup method [1].

$t = 90$ min. **Figure 4** shows the schematic diagram of Rn-222 exhalation rate measuring in the active setup method. The background value is subtracted from the measured Rn-222 activity concentration level for each sample. All samples measure in n times to get average results. By Alpha View-Expert software, final activity concentration of Rn-222 gas (A_0) is computed. Rn gas exhalation rate was computed as

$$E_x = A_0 \lambda \left(\frac{V}{F} \right) \tag{9}$$

where E_x is the Rn gas exhalation rate (Bq m^{-2} h^{-1}), A_0 is the final activity of Rn gas (Bq m^{-3}), λ is Rn decay constant (7.567×10^{-3}), V is the emanation container volume (m^3) and F is the area of each sample (m^2).

4. Level of Rn-222 in granite building materials

The level of Rn-222 activity concentrations in dwellings approximately is normal, with a trend for high concentrations to lie above those predicted by this distribution. According to the United Nations Scientific Committee on the Effects of Atomic Radiation (UNSCEAR 2000) [15], the worldwide concentration and population-weighted values of these parameters for dwellings is found to be 25 and 2.5, respectively. Rn-222 concentrations in dwellings are variable between countries because of differences in climate and geology, in techniques and construction materials and in domestic customs. The arithmetic means of radon concentration level for countries various from 12 to 140 Bq m^{-3}. Rn-222 activity concentration levels for some European countries are given in **Table 1** [18].

The rate of entry of radon from the floor and the ceiling of the reference house would amount to about 6 Bq m^{-3} h^{-1}, while the contribution from the external walls would be about 0.4 Bq m^{-3} h^{-1}. The contribution from radon exhalation from the building materials to the radon concentration in the reference house is thus estimated to be 6.4 Bq m^{-3} h^{-1}, if the air exchange rate is taken to be 1 h^{-1} [12].

Considerably greater values of the rate of entry of radon are expected to be obtained when building materials with high ^{226}Ra concentrations and normal emanating power are extensively used. Examples of such building materials are granite, Italian tuff and alum-shale lightweight concrete. Among these materials, Swedish alum-shale lightweight concrete has the highest ^{226}Ra concentrations (about 1300 Bq kg^{-1} on average) and probably the highest ^{222}Rn mass exhalation rate (440 μBq kg^{-1} s^{-1}). Assuming a density of 2000 kg m^{-3} and a diffusion length of 7.4×10^{-2} m with 0.2 m thick slab of 100 m^2 would lead to a rate of entry of radon of about 80 Bq m^{-2} h^{-1} in the reference house [16].

Techniques for reducing the radon entry rate due to exhalation from building materials have been investigated. In Sweden, aluminium foil has been applied to the walls of houses built

Country	Number of houses sampled	Period of exposure	Duration of exposure	Sample characteristics	Radon-222 concentration (average) (Bq m^{-3})
Belgium	300	1984–1990	3 months to 1 year	Population-based (selected acquaintances)	48
Czechoslovakia	1200	1982	Random grab sampling	–	140
Demark	496	1985–1986	6 months	Random	47
Finland	3074	1990–1991	1 year	Random	123
France	1548	1982–1991)	3 months (using open alpha track detectors)	Biased (not stratified)	85
Germany	7500	1978–1984 1991–1993	3 months 1 year	Random	50
Greece	73	1988	6 months	—	52
Hungary	122	1985–1987	2.5 years	Preliminary survey	55
Ireland	1259	1985–1989	6 months	Random	60
Italy	4866	1989–1994	1 year	Stratified random	75
Luxembourg	2500	1991	–	–	65
Netherlands	100	1982–1984	1 year	Random	29
Norway	7525	1987–1989	6 months	Random	60
Portugal	4200	1989–1990	1–3 months	Volunteers in a selected group (high school students)	81
Spain	2000	Winter of 1988–1989	Grab sampling	Random	86
Sweden	1360	1982–1992	3 months in heating season	Random	108
Swaziland	1540	1982–1990	3 months (mainly in winter)	Biased (not stratified)	70
United Kingdom	2093	1986–1987	1 year	Random	20.5

Table 1. Radon-222 levels in dwellings of some European countries [18].

with aerated concrete based on alum shale; the results showed a 50% reduction in the radon entry rate. In the United States, various radon sealants have been tested under conditions representative of normal construction conditions. The radon exhalation rate was reduced from 20 to 80% depending on the surface coating used. Because of the trapping of radon decay products in the wall, there is a relatively small increase in the external gamma dose rate. It was also noted that any cracks that later developed in a sealant such as paint may lead to leaks that negate a large portion of the sealing effectiveness of the paint. Similarly, the radon exhalation rate from any unpainted areas may be increased as they offer to radon a path of least resistance in comparison with painted areas. Radon-222 source with characteristics for building materials is shown in **Table 2** [12].

Material	Country	Number of sampling	Radon-222 mass exhalation rate (μBq kg^{-1} s^{-1})
Concrete			
Heavy concrete	USSR	18	3.2
Light weight concrete	USSR	19	4.1
Ordinary concrete	Sweden	3	11–31
Aerated concrete based on alum shale	Sweden	1	580
Alum-shale concrete	Denmark	1	440
Fly-ash concrete (4%)	USA	8	10
Fly-ash concrete Greece	Greece	4	6.4–20
Concrete	Hungary	100	7.8
Concrete	Denmark	4	4.7
Concrete	Norway	137	9.5–16
Concrete	Greece	–	2.9–5
Concrete	USA	50	2.5–20
(Adopted reference value)	–	–	10
Brick			
Red brick	USSR	12	1.6
Red brick	Hungary	200	3.9
Red brick	Poland	3	18
Red brick	USA	6	1
Brick	Denmark	2	0.17
Brick	Norway	18	4.2
Brick	Greece	5	0.3–7.5
Silicon brick	Poland	3	1.8
Adobe brick	USA	2	3.5
(Adopted reference value)	–	–	2
Gypsum			
Gypsum	USA	12	6.3
Gypsum board	Denmark	1	0.23
By-product gypsum (apatite)	Poland	1	2.6
Lightweight expanded			
Clay aggregate	Norway	12	5.2
Storage rock			
Storage rock	USA	9	5
Wood			
Wood	USA	2	0.2

Material	Country	Number of sampling	Radon-222 mass exhalation rate (μBq kg^{-1} s^{-1})
Sand			
Sand	USA	2	12
Sand	USA	2	3
Gravel			
Gravel	USA	4	2.2

Table 2. Radon-222 source with characteristics for building materials [12].

Acknowledgements

The author would like to thank Iran's Association of Manufacturers, which provided the original granite materials from which the samples were prepared also gave the useful information about their origin.

Author details

Akbar Abbasi

Address all correspondence to: akbar.abbasi@kyrenia.edu.tr

Faculty of Marine Sciences, University of Kyrenia, Kyrenia, North Cyprus, Mersin, Turkey

References

[1] Abbasi A, Mirekhtiary F. Comparison of active and passive methods for radon exhalation from a high-exposure building material. Radiation Protection Dosimetry. 2013:nct163; **157**(4):570–574. DOI: https://doi.org/10.1093/rpd/nct163

[2] Al-Jarallah MI, Abu-Jarad F. Determination of radon exhalation rates from tiles using active and passive techniques. Radiation Measurements. 2001;**34**(1):491–495. DOI: 10.1016/S1350-4487(01)00213-X

[3] Keller G, Hoffmann B, Feigenspan T. Radon permeability and radon exhalation of building materials. Science of the Total Environment. 2001;**272**(1):85–89. DOI: 10.1016/S0048-9697(01)00669-6

[4] Petropoulos NP, Anagnostakis MJ, Simopoulos SE. Photon attenuation, natural radioactivity content and radon exhalation rate of building materials. Journal of Environmental Radioactivity. 2002;**61**(3):257–269. DOI: 10.1016/S0265-931X(01)00132-1

[5] Righi S, Bruzzi L. Natural radioactivity and radon exhalation in building materials used in Italian dwellings. Journal of Environmental Radioactivity. 2006;**88**(2):158–170. DOI: 10.1016/j.jenvrad.2006.01.009

[6] Allen JG, et al. Assessing exposure to granite countertops – part 2: Radon. Journal of Exposure Science and Environmental Epidemiology. 2010;**20**(3):263–272. DOI: 10.1038/jes.2009.43

[7] Snelling, A., & Woodmorappe, J. Rapid Rocks—Granites... they didn't need millions of years of cooling. Creation Ex. Nihilo, 1998;**21**(1):37–39

[8] UNSCEAR, Ionizing Radiation. Sources and Biological Effects. New York: United Nations; 1982. p. 19, Supplement No. 45 (A/37/45), P-17

[9] Mustapha M, Weil D, Chardenoux S, Elias S, El-Zir E, Beckmann JS, Loiselet J, Petit C. An α-tectorin gene defect causes a newly identified autosomal recessive form of sensorineural pre-lingual non-syndromic deafness, DFNB21. Human Molecular Genetics. 1999;**8**(3): 409–412. DOI: 10.1093/hmg/8.3.409

[10] Anjos RM, Ayub JJ, Cid AS, Cardoso R, Lacerda T. External gamma-ray dose rate and radon concentration in indoor environments covered with Brazilian granites. Journal of Environmental Radioactivity. 2011;**102**(11):1055–1061. DOI: 10.1016/j.jenvrad.2011.06.001

[11] Lee SC, Kim CK, Lee DM, Kang HD. Natural radionuclides contents and radon exhalation rates in building materials used in South Korea. Radiation Protection Dosimetry. 2001;**94**(3):269–274. DOI: 10.1093/oxfordjournals.rpd.a006499

[12] UNSCEAR S. Report Sources, Effects and Risks of Ionizing Radiation-Annex B: Exposures from nuclear power production [Internet]. New York: United Nations; 1988. ANNEX A, P.68 [Accessed: 19 July 2012]

[13] Koblinger L. Mathematical models of external gamma radiation and congruence of measurements. Radiation Protection Dosimetry. 1984;**7**(1–4):227–234. DOI: 10.1093/oxfordjournals.rpd.a082998

[14] Mustapha AO, Patel JP, Rathore IV. Assessment of human exposures to natural sources of radiation in Kenya. Radiation Protection Dosimetry. 1999;**82**(4):285–292. DOI: 10.1093/oxfordjournals.rpd.a032637

[15] UNSCEAR S. Effects of Ionizing Radiation. New York: United Nations; 2000. pp. 453–487

[16] Tzortzis M, Tsertos H, Christofides S, Christodoulides G. Gamma radiation measurements and dose rates in commercially-used natural tiling rocks (granites). Journal of Environmental Radioactivity. 2003;**70**(3):223–235. DOI: 10.1016/S0265-931X(03)00106-1

[17] Asgharizadeh F, Abbasi A, Hochaghani O, Gooya ES. Natural radioactivity in granite stones used as building materials in Iran. Radiation Protection Dosimetry. 2011:ncr233; **149**(3):321–326. DOI: 10.1093/rpd/ncr233

[18] Tyson JL, Fairey PW, Withers CR. (1993). Elevated radon levels in ambient air (No. NEI-FI-222 (V. 4))

Chapter 5

Levels of Radon Activity Concentration and Gamma Dose Rate in Air of Coal Mines in Bosnia and Herzegovina

Zejnil Tresnjo, Jasmin Adrovic and Ema Hankic

Additional information is available at the end of the chapter

http://dx.doi.org/10.5772/intechopen.69903

Abstract

Radon is a unique natural element because it is an inert gas and at the same time radioactive in all of its isotopes. It is known fact that exposure of the population to high concentrations of radon gas leads to irradiation of primarily respiratory organs which can cause lung cancer. Radon is a subject of intense research around the world in order to, among other reasons, assess the risk of exposure and develop appropriate standards of protection and its control. By coal mining and exploitation, radioactive radon gas, which is captured in natural geological structures, is relocated from the deep coal layers. Hence, it is concentrated in the depots and coal seams of the mines or being transported to the surface of the earth where it can significantly change the levels of radioactivity in the working premises and residences. This chapter presents the results of a 3-year research of radon activity concentration and gamma dose rate in the air in underground and surface coal mines of Bosnia and Herzegovina.

Keywords: radon activity concentration, gamma dose rate, air, coal mines, Bosnia and Herzegovina

1. Introduction

Coal, as well as majority of natural materials, contains natural radioactive elements [1]. The concentrations of natural radionuclides in coal are generally lower than those in the earth's crust; however, sometimes higher concentrations of uranium and thorium and their decay products in different layers of coal can be found, as a result of leaching of uranium and thorium from uranium- and thorium-rich volcanic rocks.

In many mines, the presence of radioactive materials such as uranium and thorium in the surrounding rocks may lead to increased concentrations of the isotopes of radon and to an increase of radon decay products produced in the mine atmosphere. The atmosphere in the underground mines contains, in gaseous form or in form of aerosols, all natural chain of radionuclides originated by the decomposition of thorium and uranium. As a result, the concentration of certain radionuclides, especially radon and its decay products, can achieve high levels of concentration which can be dangerous to workers' health [2].

Coal mining and its combustion in thermal power plants is potentially the most important process for creating technological elevated levels of natural radioactivity. Opening of large surface mines leads to accelerated emanation of radon from the deeper layers of the earth's crust [3]. Similarly, in coal mines, increased accumulation of radon can be caused due to its inflow through cracks and fractures in land and particular via the underground water systems.

Dangers caused by radon and its decay products are identified in the seventeenth century and in the last century, radon is considered responsible for the increased incidence of lung cancer among miners who were exposed to radon. For this reason, there was a need for atmospheric research in underground mines, as well as for dosimetry monitoring of miners involved in these activities. Dosimetry monitoring in the coal mines is needed to assure that the activities in mines do not generate radon in quantities that exceed the limit values defined by the appropriate regulation.

Although there is no any information on the emission of radon from coal mines (surface and underground mine), the UNCSEAR Committee has judged that annually 30–800 TBq of radon is released from mines around the planet, resulting in annual collective effective doses from 0.5 to 10 Sv per man, respectively. In the report UNSCEAR 2000, the Committee estimated that the average concentration of ^{40}K, ^{238}U, and ^{232}Th in coal is 50, 20, and 20 Bq/kg, respectively. This estimate is based on the analysis of coal samples from 15 countries, and especially important is to mention that the value of certain radionuclides varies up to two orders of magnitude.

Bosnia and Herzegovina (B&H) is rich in coal, which represents one of its most important energy sources. The total geological reserves of coal in B&H are estimated at 5.647 billion tons, of which 2.540 billion tons are balance; 1.437 billion tons of lignite and 1.103 billion tons of brown coal. In coal mines in Bosnia and Herzegovina in 2015, approximately 9.2 million tons of coal is produced, of which approximately 85% of total coal production have been taken by power plants. Thus, the coal industry in Bosnia and Herzegovina is a potential source of danger of radioactive contamination of the environment, especially in cases when certain coals are used in power plants which indicate an increased content of natural radionuclides.

This chapter presents the results of 3-year research of radon activity concentration and gamma radiation in underground and surface coal mines in Bosnia and Herzegovina.

2. Coal mines in Bosnia and Herzegovina and general characteristics of coal

Economically, important coal deposits were formed at different stages of historical and geological development of the Earth and on large number of sites (more than 100). Brown coal deposits were formed mainly in the lower and middle Miocene and lignite in the upper Miocene and lower Pliocene. Weather and creation conditions of economically important coal deposits determined their rank and regional deployment. Coal basins in B&H are long been known and geologically largely explored. In some of them, mining has been done for over 100 years. The major reserves of brown coal are located in the following basins: Central Bosnia (deposits: Kakanj, Zenica, Breza, Bila), Banovici (Seona, central basin, Djurdjevik), Ugljevik (Bogutovo Selo, Ugljevik-Istok, Glinje). The most important lignite reserves are located in the basins of Kreka, Gacko, Stanari, Bugojno, Livno, and Duvno.

Coal mining in B&H coal mines is being done in two ways: surface mining and underground mining (pits). Mine sites are located directly at the bearings. **Figure 1** provides a geographical presentation of active coal mines in B&H, and in **Table 1**, general characteristics of the mines which were the subject of research are shown.

Figure 1. Position of coal mines and thermal power plants in Bosnia and Herzegovina.

Name of coal mine	Type of coal	Shelf	Top
Surface exploitation			
PK "Cubrici" Mine "Banovici"	Brown coal	Sand, marl and freshwater limestone	Gray-yellow and gray marl, marly limestone and clay
PK "Dimnjace" Mine "Gracanica"	Lignite	Sandy clay and limestone marl	Sandy clay, sandy marl
PK "Drage" Mine "Tusnica"	Lignite	Clay, clay marl	Sandstones and conglomerates
PK "Vrtliste" Mine "Kakanj"	Brown coal	Marl-sand clay	Sandstones and different conglomerates
Underground exploitation			
Pit "Glavni sloj" Mine "Mramor"	Lignite	Sand	Marl clay
Pit "Djurdjevik"	Brown coal	Sandstone, clay-marl clay	Marl, marly limestones
Pit "Haljinici" –Mine "Kakanj"	Brown coal	Marl-sand clay	Sandstones and different conglomerates
Pit "Stara Jama" Mine "Zenica"	Brown coal	Marl clay, clayey marl	Layered limestone
Pit "Raspotocje" Mine "Zenica"	Brown coal	Marl clay, clayey marl	Layered limestone

Table 1. General characteristics of coal mines in Bosnia and Herzegovina.

The coal industry in Bosnia and Herzegovina is a potential source of danger of radioactive contamination of the environment, especially in cases when certain coals are used in power plants which indicate an increased content of natural radionuclides. For these reasons, in recent years, in accordance with the relevant regulations, the Public enterprise Elektroprivreda of Bosnia and Herzegovina composed of TPP Kakanj and TPP Tuzla carried out annual level monitoring of radioactivity in the two thermal power plants. Monitoring included checking the level of radionuclides in coal used for power generation, the content of radionuclides in ash and slag, and measured levels of radon and gamma radiation at locations in the vicinity of thermal power plants.

Table 2 presents the results of measurements (mean value) of specific activities of natural radionuclides in coal mines which were the subject of these investigations.

In **Figure 2**, a comparative graphical representation of the average specific activity of radionuclides in coals from different mines is provided.

Highest specific activity of natural radionuclides, uranium and radium was measured in the coal mine Tusnica—Livno with a value of 26.68 ± 1.07 Bq/kg for U-235, 623.03 ± 23.32 Bq/kg for U-238, and 1191.34 ± 4.83 Bq/kg for R-226. The highest values of Th-232 are measured in the coal mine Tusnica—Livno with value of 26.67 ± 1.86 Bq/kg, whereas the highest values of potassium in coal mines were measured in mine Vrtliste—Kakanj. If it is known that, under appropriate regulation, the maximum activity for R-226 is 400 Bq/kg or 300 Bq/kg for U-238;

Coal mine	Activity (Bq/kg)				
	U-235	U-238	Th-232	Ra-226	K-40
Vrtliste—Kakanj	1.82 ± 0.21	39.58 ± 6.63	21.78 ± 1.00	42.05 ± 1.24	199.09 ± 6.84
Haljinici—Kakanj	2.54 ± 0.31	55.08 ± 6.64	17.69 ± 0.81	46.62 ± 1.27	129.57 ± 5.82
Tusnica—Livno	26.68 ± 1.07	623.03 ± 23.32	26.67 ± 1.86	1191.34 ± 4.83	33.92 ± 7.20
Zenica	1.08 ± 0.18	23.36 ± 3.84	5.61 ± 0.47	23.44 ± 0.71	40.68 ± 3.39
Gracanica—G. Vakuf	7.62 ± 0.86	165.41 ± 10.07	4.25 ± 0.63	145.14 ± 1.74	16.92 ± 1.86
Banovici	1.67 ± 0.28	36.17 ± 6.05	16.50 ± 0.76	48.40 ± 1.79	76.30 ± 3.28
Mramor-Kreka	1.75 ± 0.34	38.09 ± 7.35	11.93 ± 0.57	48.52 ± 1.77	71.67 ± 3.08
Djurdjevik	2.58 ± 0.40	56.00 ± 8.69	9.03 ± 0.44	26.51 ± 1.02	67.37 ± 17.92

Table 2. Specific activity of natural radionuclides in coal [4].

then from the **Table 2**, it can be concluded that brown coal from Tusnica has significantly increased levels of natural radionuclides in these two elements (2–3 times). Because of this knowledge, the percentage of coal from the mine Tusnica in electricity production in TPP Kakanj has been reduced, and thus the total level of natural radionuclides in coal used by the power plant is below the maximum levels.

Comparing these values with average values of radionuclides in coals of some European countries (UNSCEAR report 2000), it can also be concluded that the brown coal from the mine "Tusnica" has a significantly elevated level of natural radionuclides. These results of gamma spectrometry analysis were the main reason for conducting thorough research activity of concentration level of radon and gamma dose rate in coal mines in Bosnia and Herzegovina.

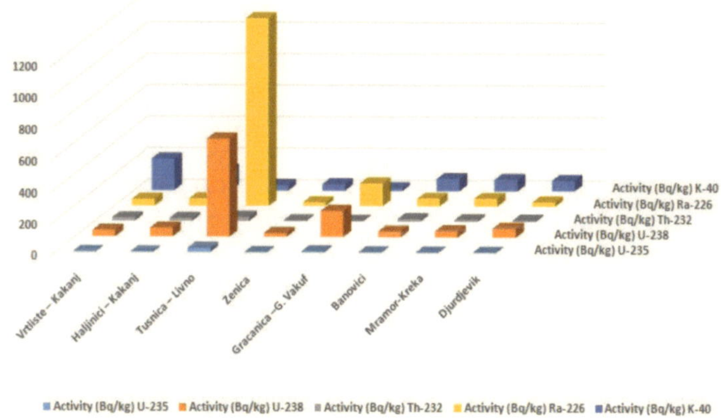

Figure 2. The mean specific activity of radionuclides in coals.

3. Research methods and instrumentalization

For creation of the experimental part of this work, the most advanced and most modern research methods of detection and dosimetry of ionizing radiation were used: a method of solid-state nuclear track detector (SSNTD) and method of measuring radiation ionization chamber. In addition to these measurements on the test sites, measuring of gamma radiation dosage was also conducted.

Measurement of radon concentration using nuclear track detectors was performed with Radon system (Radosys, manufactured in Hungary). The basic components of this system are diffuse dosimeter with the detector type CR-39; system for chemical analysis of detectors; and system for automatic reading of detectors. We used detectors made of poly-allyl-diglycol-carbonate, known as CR-39, with dimensions 10 × 10 × 1 mm, sensitivity to alpha tracks of 2.9 tracks per (cm^3 kBq h)/m^3. Limit saturation is greater than 12,000 kBqh/m^3. This method is based on registration of traces of particles from the radioactive decay of radon in the dielectric detector. Detection of charged particles using these detectors is performed with the tracks of the particles into which they penetrate the detector (latent traces). With this method, radon activity concentration was measured exclusively in the mine atmosphere and the measurement of the concentration of its decay products was avoided. **Figure 3** schematically shows the process of measuring radon using "RadoSys" system.

Error in the measurement of radon activity concentration by nuclear track detectors is calculated according to formula below:

$$\sigma = A\sqrt{\sigma_\rho^2 + \sigma_K^2 + \sigma_\tau^2} \tag{1}$$

where A is the radon activity concentration,

$$\sigma_\rho = \frac{1}{\sqrt{N}} + 0.004$$

where N is the number of traces of α particle in a detector, σ_K is the calibration error coefficient (for CR-39 $\sigma_K = 0.18$), and σ_τ is error due to transport of dosimeter: $\sigma_\tau = \frac{\text{duration of transport}}{\text{duration of exposure}}$.

For the second method, which was done as a comparative research method, measuring system AlphaGUARD PQ 2000 PRO/MC503 was used (Genitron Instruments, made in Germany), whose work methodology is based on the principle of ionization [5]. The device can operate in two operating modes: diffusion and pump mode. In diffusion mode, radon diffuses into the ionization chamber through a glass fiber filter that prevents the entry of radon decomposition products or aerosols while for the pumping regime, "AlphaPUMP" is used. Measuring range of radon activity concentration of this product is 2–2 × 10^6 Bq/m^3, while the temperature range of −10 to 50°C. Calibration error of ^{222}Rn is 3%. AlphaGUARD cylindrical ionization chamber device has an active volume of 0.56 dm^3. The basic configuration of this system are AlphaGUARD Radon Monitor and "Data Expert" software packages (**Figure 4**).

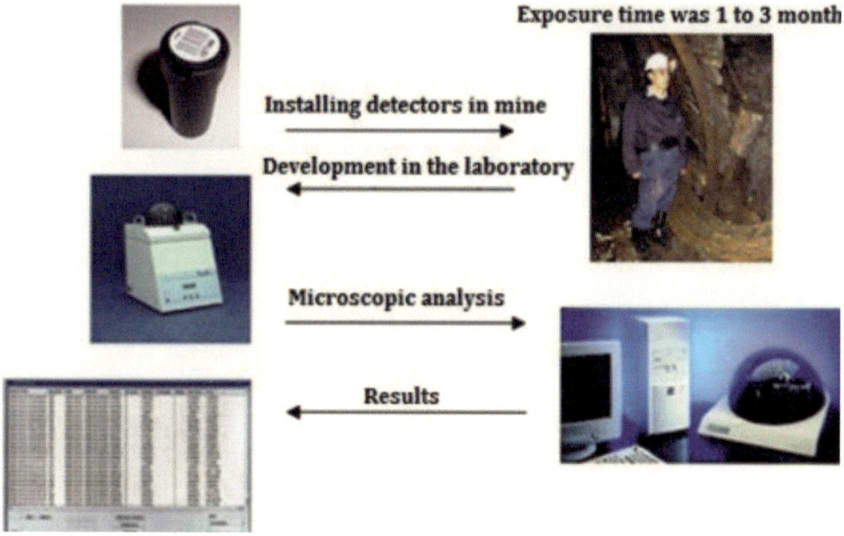

Figure 3. Order of measurement processes of radon in mines using "RadoSys" system.

To measure the dosage of gamma radiation in the mine atmosphere and at the location around the mine, ADL Gamma Tracer system was used, the latest product of Genitron Instruments GmbH, Frankfurt, Germany. Complete electronics and power supply are located in hermetic housing. The basic technical characteristics of ADL Gamma Tracer are: Detector Geiger-Muller tube, the measurement range of 10–10 Sv/h, the energy dependence of 45–1300 keV, the sensitivity is 2 × 0.2 counts/s for 100 nGy/h; the measurement cycle can be set to 1, 2, 5, 15, 30, 60, and 120 min (specified by user); data capacity is 12,800 measurement points (4–1060 days). Through interactive infrared port, the registered values could be dislocated at any time. Gama professional software program for the usable communication and as a software for analyzing, guarantees an easy, fast, and secure access to the accumulated data and their visualization. In **Figure 5**, the most important components of Gamma Tracer system are given.

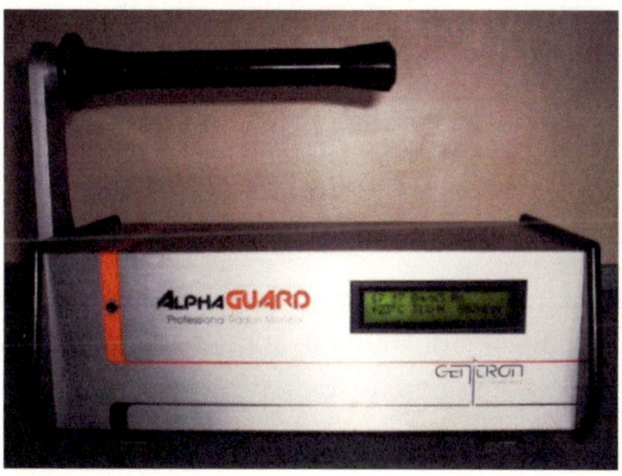

Figure 4. AlphaGUARD PQ 2000 PRO (manufacturer: Genitron Instruments).

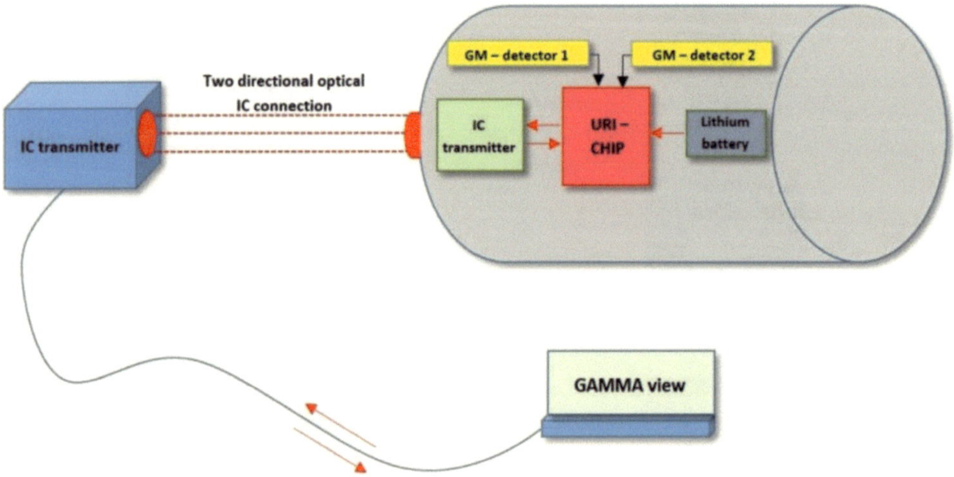

Figure 5. Basic components of Gamma Tracer system.

4. Results and discussion

In order to better and clearer perceive the contribution of natural radiation from technologically modified natural radioactivity, measurements of activity concentrations of radon were carried out, in most cases, during the most intensive work at surface of coal mines, when the layers and interlayers of coal and tailings on the surface mines were opened, as well as in all the activities that are performed daily in the underground coal mining. In addition to the mine site, for comparison, studies were also carried out in the immediate vicinity, as well as in the smaller (and sometimes larger) urban core.

Measurement sites were chosen after extensive analysis of all parameters that would be of importance for the research results. Also, for all measuring points on all test locations, following items have been taken into account:

- Setting of dosimeters was conducted immediately after their assembling, and thus reducing the effect of background values of radon.
- During transport, dosimeters were protected with aluminum foil.
- Reading of dosimeters was performed at the Laboratory for Dosimetry and Radiation Protection (LDDZZ) at Faculty of Sciences of the University of Tuzla.
- Measurement of radon concentration using an AlphaGUARD PQ 2000 PRO was done twice, the first time while placing dosimeters at the test site and the second time while taking the dosimeter after completing measurement.
- All measurements of radon concentration in air using an AlphaGUARD PQ 2000 PRO were carried out in the time interval of 10 min.

The research results are presented in **Tables 3–5**.

Name of coal mine	Number of measurement locations	RAC (Bq/m³)		
		Minimum	Maximum	Average
Surface exploitation				
PK "Cubrici"	10	5 ± 2	44 ± 9	22 ± 5
PK "Dimnjace"	16	12 ± 4	45 ± 10	21 ± 6
PK "Drage"	10	19 ± 5	50 ± 10	24 ± 6
PK "Vrtliste"	10	7 ± 3	36 ± 8	20 ± 5
Underground exploitation				
Pit "Glavni sloj"	30	7 ± 2	71 ± 14	32 ± 7
Pit "Djurdjevik"	20	25 ± 6	76 ± 16	52 ± 11
Pit "Haljinici"	10	13 ± 4	50 ± 11	27 ± 7
Pit "Stara Jama"	20	910 ± 3	106 ± 21	27 ± 7
Pit "Raspotocje"	20	12 ± 3	38 ± 8	20 ± 5

Table 3. Results of RAC measurements at the workplaces of coal mines in B&H using nuclear track detectors.

Name of coal mine	Number of measurement locations	RAC (Bq/m³)		
		Minimum	Maximum	Average
Surface exploitation				
PK "Cubrici"	10	5	111	35
PK "Dimnjace"	16	15	55	32
PK "Drage"	10	26	60	38
PK "Vrtliste"	10	14	33	21
Underground exploitation				
Pit "Glavni sloj"	30	12	92	43.5
Pit "Djurdjevik"	20	5	3420	279.5
Pit "Haljinici"	10	24	835	82.5
Pit "Stara Jama"	20	5	78	34.5
Pit "Raspotocje"	20	9	32	17.5

Table 4. Results of RAC measurements at the workplaces of coal mines in B&H using AlphaGUARD PQ 2000 PRO device.

As it can be seen from **Table 3**, the maximum mean concentration of radon (RAC) was measured using nuclear track detectors on the surface pit at PK "Drage" of coal mine Tusnica—Livno and the value is 24 ± 6 Bq/m³.

Name of coal mine	Number of measurement locations	Gamma radiation equivalent dose rates in air (nSv/h)		
		Minimum	Maximum	Average
Surface exploitation				
PK "Cubrici"	10	44	254	134.0
PK "Dimnjace"	16	73	108	87.0
PK "Drage"	10	80	106	91.0
PK "Vrtliste"	10	68	112	90.0
Underground exploitation				
Pit "Glavni sloj"	30	38	125	72.0
Pit "Djurdjevik"	20	48	180	97.5
Pit "Haljinici"	10	56	184	97.0
Pit "Stara Jama"	20	24	152	83.0
Pit "Raspotocje"	20	45	109	89.5

Table 5. Gamma radiation equivalent dose rates in working places of coal mines in Bosnia and Herzegovina.

No.	Site	RAC (Bq/m^3)
1	Entrance to the old mining-1	27 ± 6
2	Entrance to the old mining-2	29 ± 6
3	Separation-1	25 ± 6
4	Separation-2	26 ± 6
5	Old warehouse Martinovac	19 ± 5
6	Warehouse of administration building	46 ± 10
7	Entrance desk (entrance to the mine)	44 ± 9
8	Spring Mandek-Novakovac. Water intake for water supply Tusnica	50 ± 10
9	Near place (about 200 m from mine)	45 ± 9
10	Administration building "Mine a coal Tusnica" in Livno-cellar	73 ±14
11	Administration building "Mine a coal Tusnica" in Livno—first floor	43 ± 9
12	Livno—periphery (private house)	57 ±11

Table 6. Measurement results of radon activity concentration obtained using nuclear track detectors at pit PK Drage (near and far surrounding).

If we take into account all measurement sites at the locality of PK Drage (**Table 6**), it can be seen that the maximum concentration of radon in the air at this location was 50 ± 10 Bq/m^3. Among other reasons, the reason for such a huge concentration of radon can be found in the fact that coal from this mine contains the largest percentage of radionuclides of uranium and radium with whose disintegration occurs radioactive gas radon, which is transported by diffusion from the depths to the surface. Regardless of this, the value of radon concentration in the air at the mine surface and at other mines are slightly higher than usual, which can be measured in most European countries, ranging from 10 to 20 Bq/m^3.

Short-term measurements of RAC using AlphaGUARD PQ 2000 PRO (ionization chamber) conducted at this open pit show some greater value than those obtained with nuclear track detectors (**Table 4**), which can be attributed with variety of weather conditions prevailing at the time of measurement (temperature, pressure, humidity, speed of moving air, etc.). The maximum value of measured radon activity concentration in the PK ''Cubrici'' was 111 Bq/m3, while the highest mean radon activity concentration was measured in the PK ''Drage.'' In **Figure 6**, a comparative graphical representation of the means of RAC for opencast mining measured by these methods is presented.

Table 5 shows that the mean value of equivalent dose strength of gamma radiation in the working environment in open pits PK "Cubrici" is 134 nSv/h or 1.174 mSv/y, which is less than the global effective annual dose of natural resources. On the basis of UNSCEAR-1993rd results, the effective equivalent dose of average radon concentration is 40 Bq/m^3 in buildings and 10 Bq/m^3 in an open space, and for its decay products is 1.2 mSv/y and 1 mSv/y for other natural radioactive sources. In **Figure 7**, a graph of the mean values of intensity of an equivalent dose of gamma radiation in the working environment of open pits is given.

Based on the obtained results of measured radon concentration in coal mines (pits) B&H, which are given in **Table 3**, it can be concluded that concentration is insignificant. The highest average value of RAC measured by nuclear track detectors was 52 ± 11 Bq/m^3 in the pit "Djurdjevik," whereas the maximum RAC measured at one measurement location was 106 ± 21 Bq/m^3 in the pit "Stara Jama" of brown coal mine "Zenica." Although the average value of RAC in the pit "Djurdjevik" is significantly higher compared to other

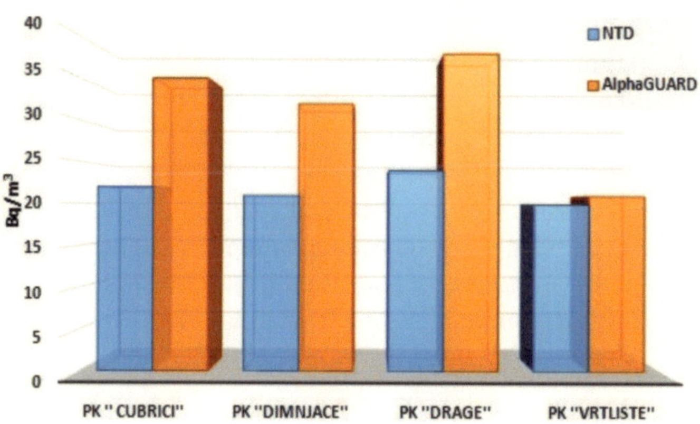

Figure 6. Graphic presentation of medium value of RAC obtained using both nuclear methods.

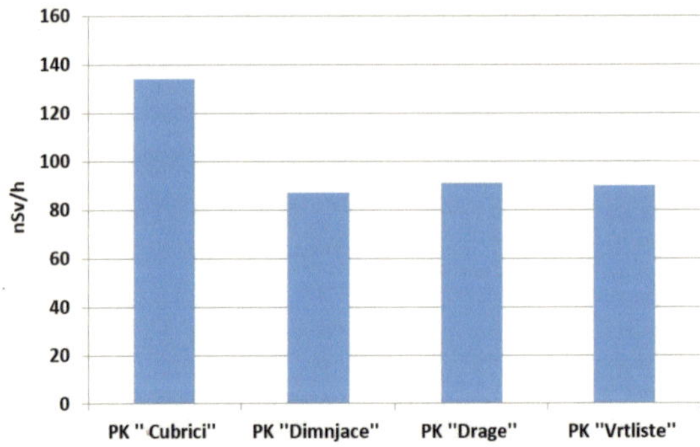

Figure 7. Graphic presentation of mean value of the equivalent dose rate of gamma radiation in working environment of open pits.

pits (somewhere twice), but in general it can be concluded that all these values are below the levels which are measured in the coal mines of some European countries. Causes of an increased concentration of radon in the air of pit "Djurdjevik" need to be sought in the geological structure of the tops and shelves of mines, and large quantities of underground water which were present in the pit during the measurements. Geology of the pit shelf varies from place to place, usually there are conglomerates, clay-marl clay, carbonaceous clay, while the tops of the pits consist of thick series of marls and marly limestone (**Table 1**). Having in mind that a very compact crystal structures represent an ambience with small penetration of radon gas (which is not the case with sand pits and rocks with cracks and fractures) and that a thick layer of clay is not only badly leaking the gas radon but also can change the direction of movement, then the values of RAC in the air of the cave "Djurdjevik" have their justification.

Measurements of RAC using an AlphaGUARD PQ 2000 PRO in the pits and surface mines also show something of greater values than those obtained with nuclear track detectors (**Figure 8**). These results variations can be attributed not only to different meteorological conditions prevailing at the time of measurements (temperature, pressure, humidity, speed of moving air, etc.) but also to the geological characteristics and mining operations during the measurement (drilling, blasting, dredging, transport, etc.).

The maximum RAC (**Figure 9**) measured by this method was 3420 Bq/m^3 and had been measured in the pit "Djurdjevik," on the measuring site which was located near to the work site (coal excavation). This current value of increased RAC can be explained that during the excavation, there was a sudden release of radon that was trapped in the coal seams, which resulted in increasing the current RAC. That this was a current value of RAC, it was proved with repeated measurements with both methods of measurement. The same case happened in the pit "Haljinici" where at one measuring site measurement shows value of 835 Bq/m^3. Overall, the largest mean RAC was measured in the pit "Djurdjevik" and amounted to 279.5 Bq/m^3.

The first measurement of radon using an AlphaGUARD PQ 2000 PRO in the pit "Djurdjevik" and in its immediate vicinity was conducted in November. When on some locations obtained

http://dx.doi.org/10.5772/intechopen.69903

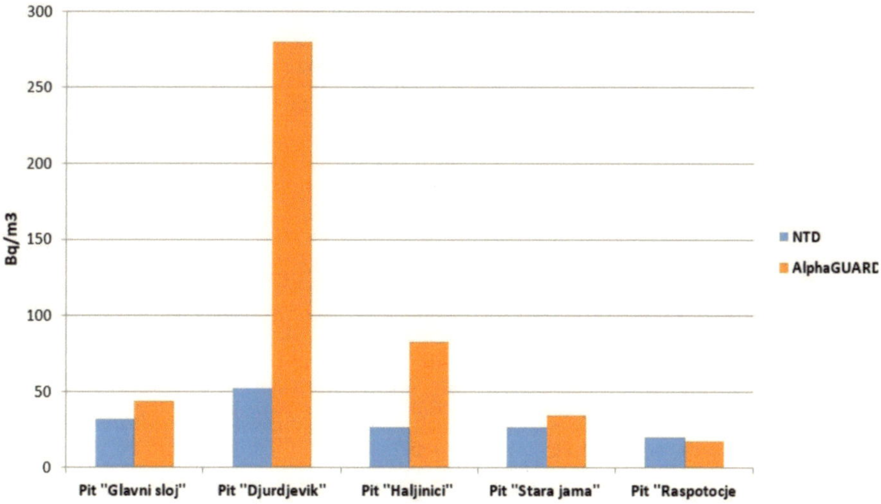

Figure 8. Graphic presentation of RAC values in the mine pits in BiH.

values of RAC were high, additional measurements were conducted at the same locations in April. The measurement results show that the current values of RAC can vary considerably from one location to another and can range from 48 to 3420 Bq/m^3 at the first measurement and from 5 to 475 Bq/m^3 in the second measurement, which can best be seen in **Table 7**.

Figure 10 shows a graph representation of the mean rate of the equivalent dose of gamma radiation in the air of coal mines with pit exploitation. The highest mean rate of the equivalent dose of gamma radiation was detected in the pit "Djurdjevik" and amounted 97.5 nSv/h, which is also less than the global annual effective dose from natural sources.

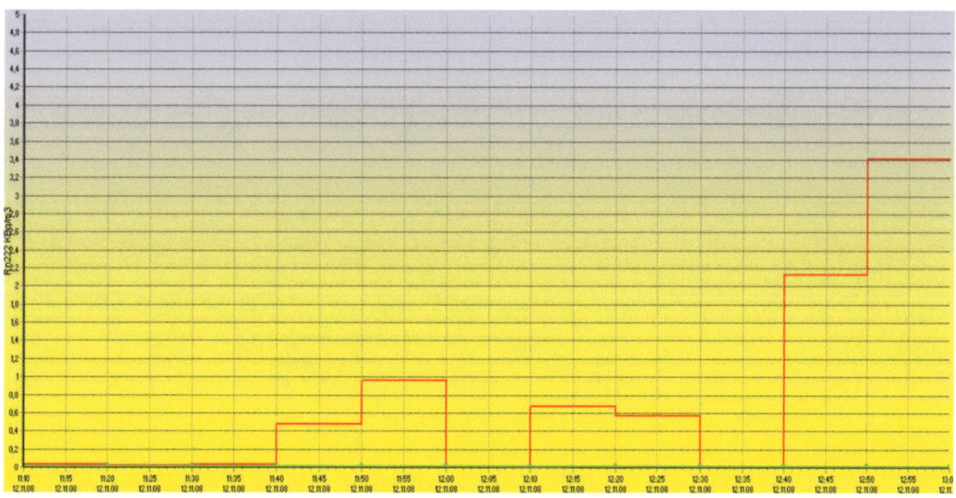

Figure 9. Graphic presentation of RAC [Bq/m^3] in the air of pit "Djurdjevik".

No.	Location (description)	RAC (Bq/m³)	
		First measurement	Second measurement
1	GDN 1	60	260
2	Level 180	48	475
3	GDN – Level 165	60	5
4	GDN – Level 137	60	5
5	GDN – Level 106	480	5
6	GDN – Level 90	770	5
7	Kota 55	48	12
8	Intersection GDN DH 25	670	12
9	Intersection TU 35 – DH 25	670	25
10	DH 25 between TU 35 and VU 6	580	25
11	VH 6 – on top	54	25
12	TH 18 – Intersection TN – (-2)	54	25
13	Explosion storage number 8 – TM (-2)	2340	55
14	Work site 1 – Preparation 36	3420	55
15	TH 18 – Intersection DP 18	220	40
16	TH 18 – Intersection IP 20	220	40
17	Intersection DNA – 2	54	46
18	Chamber tape – 2	54	38
19	TH 106 – by Vitla	60	18
20	GTN – entrance to the pit	60	30

Table 7. The values of radon concentration in the air of pit "Djurdjevik" obtained by AlphaGUARD PQ 2000 PRO device.

The calculation of correlation factor, r_{xy} is performed according to known statistical formula [6]:

$$r_{xy} = \frac{\sum_i x_i y_i - \frac{1}{n}\sum_i x_i \sum_i y_i}{\sqrt{\left[\sum_i x_i^2 - \frac{1}{n}\left(\sum_i x_i\right)^2\right]\left[\sum_i y_i^2 - \frac{1}{n}\left(\sum_i y_i\right)^2\right]}} \tag{2}$$

which enables further analysis of the obtained results. A significant correlation exists if the ratio is $r_{xy} \geq 0.5$. By calculating a linear correlation coefficient r_{xy} between the radon activity concentrations measured by nuclear track detectors in two different time periods (spring and autumn) in the mine Djurdjevik, considerable correlation between the two measurements is observed

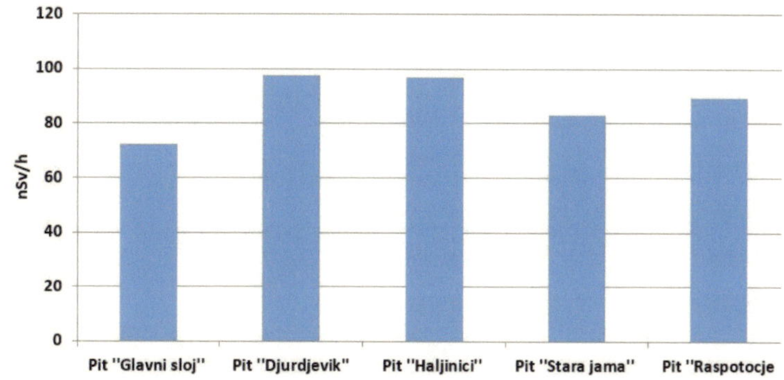

Figure 10. Graphic presentation of mean value of the equivalent dose rate of gamma radiation in workplace pit mine a B&H.

(r_{xy} = 0.64). In the same mine, significant correlation was also observed between the radon activity concentration measured using AlphaGUARD device and the equivalent dose strength of gamma radiation measured in spring (r_{xy} = 0.64), while the poor correlation is revealed by the first series of measurements conducted in autumn (r_{xy} = 0, 11), which can be attributed to current weather conditions (temperature, pressure, humidity, air speed moving, etc.). A similar correlation ratio was observed with the results of measurements on the pit surface "Drage" of coal mine "Tusnica."

5. Calculation of effective dose

Assessment of the mean annual effective dose for workers who work on the surface and pit mines due to exposure to radon and its decomposition products can be calculated using a simple model (Annex B of the report UNSCEAR 2000) [3].

$$E = k_1 T C_{Rn} + k_2 F T C_{Rn} = (k_1 + k_2 F) T C_{Rn} \tag{3}$$

wherein, k_1 = 0.17 nSv/(Bq/m^3 h) and k_2 = 9.0 nSv/(Bq/m^3 h) are the conversion factors, T is the residence time in hours per year in the atmosphere with the concentration of radon CR, and F is the factor of balance.

Taking into account the results of measurement of RAC in the previous tables, and having in mind that a worker at the workplace on an average spends 8 hours a day (2920 hours per year), and using that F = 0.4 for indoors and F = 0.6 for outdoors, medium effective dose for workers in coal mines in B&H can be assessed. The results of these calculations are given in **Table 8**.

According to the UNSCEAR 2000 report, the average radon concentration amounts 40 Bq/m^3 for indoors, and 10 Bq/m^3 for outdoors. Substituting these values in the relation (3) provided that the values of other parameters are unchanged, the average annual effective dose originating from radon and its decay products in an enclosed space is obtained and amounts 0.440 mSv/y,

	Coal mine	Effective dose (mSv)	
		Interval	Average
1	Cubrici	0.055–0.484	0.242
2	Dimnjace	0.132–0.495	0.231
3	Drage	0.209–0.550	0.264
4	Vrtliste	0.077–0.396	0.220
5	Glavni sloj	0.077–0.782	0.352
6	Djurdjevik	0.275–0.837	0.572
7	Haljinici	0.143–0.550	0.297
8	Stara Jama	0.110–1.167	0.297
9	Raspotocje	0.132–0.418	0.220

Table 8. Values of the annual effective dose for workers on the surface and pit mines as a result of exposure to radon and its decomposition products (nuclear track detector).

respectively 0.163 mSv/y for indoors. Comparing these values to the values from **Table 8** it is seen that the workers of mine "Djurdjevik" were exposed to somewhat higher dose than the usual, and to almost twice higher dose in the mine area where drilling and excavation are carried out, where we measured the maximum value.

Assessed values of the effective dose of employees at surface pit "Drage" of coal mine "Tusnica," as a result of exposure to radon and its decomposition products are ranged from 0.209 to 0.550 mSv/y, with average value of 0.264 nSv/y.

6. Conclusions

During digging and coal mining, radioactive radon gas, which is contained in natural geological structures, is being redistributed from the depth of coal beds. Thus, this gas can be concentrated in depots and stopes of mines or can be transported to the surface of the earth, which can significantly change the levels of radioactivity and radioecology picture in the working and living spaces. This chapter presents the research results of activity concentration level of radon and gamma radiation in the air in a coal mine in Bosnia and Herzegovina, as on the surface coal mines and in the pits, obtained with most contemporary measuring devices. On the basis of all knowledge about radon and its decomposition products, a methodology was defined and research was done in mines with nuclear measuring methods which are worldwide mostly used for measuring radon in mines: nuclear track detectors method (passive method), a method of measuring radiation with ionization chamber (control method). At the same locations in the mine atmosphere, simultaneous measurement of intensity of gamma radiation using an "ADL Gamma Tracer" system is performed. Based on the obtained results, assessment of the mean annual effective dose for workers who work in underground mines as a result of exposure to radon and its decomposition products was carried out.

With respect to the property of radon that it dissolves very well in water, influence of the amount of precipitation, intensity and underground waters to the levels of radon in the mine was noticed. Studies carried out in this chapter show that there are variations in the concentration of radon gas structure of coal, in the atmosphere and in other ambient media correlated to the characteristics of geologic structure, in function of the technology of obtaining coal and weather and climate changes. During the tests, with special attention, currently installed ventilation system at pit coal mining, as well as the presence of groundwater, was considered.

Mean values of radon activity concentration measured by nuclear track detectors in mines where surface exploitation is done are insignificant. The highest mean value of RAC obtained for PK "Drage" of coal mine "Tusnica"—Livno was 24 ± 6 Bq/m^3. The main reason why this mine showed the highest value of RAC is that the coal from this mine contains the largest percentage of radionuclides of uranium and radium with whose break-up radioactive gas radon is produced.

Preliminary measurements of RAC in housing and other facilities that are located in the immediate vicinity of open pit mines show that the mean values of RAC are ranged from 33 ± 7 Bq/m^3 in the closer environment of the mine "Banovici" to 50 ± 11 Bq/m^3 in the closer environment of surface pit "Dimnjace" of coal mine "Gracanica," are bellow the limits established by international organizations.

Short-term measurements of RAC performed using an AlphaGUARD PQ 2000 PRO device showed somewhat higher values than those obtained with nuclear track detectors on all surface mines, which can mostly be attributed to the different meteorological conditions during measurements (temperature, pressure, humidity, speed of movement of air, etc.).

On the basis of obtained results of RAC measured using nuclear track detectors in the pits of coal mines in Bosnia and Herzegovina, it could be concluded that the highest mean RAC measured in the pit "Djurdjevik" was 52 ± 11 Bq/m^3, which is below the values measured in some mines of European countries. Geology structure of rocks (sandstones, clay-marl clay, carbonaceous clay) and shelves (marls and marly limestone) of pit "Djurdjevik," as well as large quantities of underground waters (which were present in the pit during the measurement), are the main reason why the highest mean values were obtained from all other considered pits. The maximum value of RAC measured by this method is registered in the pit "Stara Jama" of coal mine "Zenica" and amounted 106 ± 21 Bq/m^3. The location where the maximum value was obtained was a location where a collision of input (fresh) and output (exhausted) air currents happens and where a high humidity in the air was registered.

Activity concentrations of radon measured using an AlphaGUARD PQ 2000 PRO device in the pits show that the current values can vary significantly from one location to another, which is particularly characteristic for the pit "Djurdjevik" (48 to 3420 Bq/m^3) and for the pit "Haljinici" of coal mine "Kakanj" (25 to 835 Bq/m^3). Proof that these results are presenting current values were repeated measurements which didn't register significant values. Such large variations of RAC in some pits are proof that coal excavation can suddenly release trapped radon in coal layers, and as a results are current increases of RAC. It is impossible to accurately identify and even predict when it will come to a sudden release of radon during coal excavation, but the measurements carried out in this chapter showed that it really can happen.

Based on the research results of RAC in the pit coal mines of Bosnia and Herzegovina, it is possible to determine the efficiency of ventilation systems for each pit separately. It has been established that currently installed ventilation system does not allow the retention of radon gas in the pit, but using it, radon is being removed from the pit and thus can increase the RAC out of the pit, and thus affect the overall radioactivity of the environment. Measurement of RAC can be used as a good indicator of the quality of the ventilation systems in the pit of the mine.

In any coal mine, radon concentration dependency of the depth is not noticed, suggesting that radon does not come from the earth depth. In fact, RAC dependence of the rocks composition that is in the environment of set dosimeters was noticed. Radon concentrations are higher where the surroundings are slates, which have a higher content of uranium.

The measured intensity dose of gamma radiation in the working environment of surface and pit mining operations using autonomous device ADL Gamma Tracer system show a slight increase compared to the level of natural background from an unspoiled natural environment, and in general it can be said that they are within the limits of normal values measured and many European countries. Measurements clearly show that there are differences between levels of the undisturbed natural environment radiation from selected measurement locations that were served for comparison and the level of natural radioactivity in the impact zone of mining operations at the considered coal mines. These differences in values of gamma radiation dosage are lower than 10%.

On the basis of all conducted research in this chapter, it is evident that every mine, every house, and every residential or other object has its own specific "life" of radon. Therefore, it is not enough to measure the RAC in only one lignite or brown coal mine and on that basis make a general conclusion, it is necessary, with a number of aspects, investigate each mine, and only then make a conclusion that is peculiar to it alone.

The estimated values of annual effective doses for employees on most workplaces at considered mines as due to exposure to radon and its decomposition products for 8 hours of work per day (2920 hours per year) were within normal values.

Having in mind the conditions in which the researches were done, especially when it comes to measurements carried out in the pits, it can be concluded that the achieved knowledge and the results obtained in this study represent a significant contribution to the creation of the initial database for mapping and forecasting geogenetic potential of radon in coal bays of Bosnia and Herzegovina, as well as creating maps of radon and natural radioactivity in Bosnia and Herzegovina.

http://dx.doi.org/10.5772/intechopen.69903

Author details

Zejnil Tresnjo[1]*, Jasmin Adrovic[2] and Ema Hankic[2]

*Address all correspondence to: zejniltresnjo@yahoo.com

1 High School "Logos Centar", Mostar, Bosnia and Herzegovina

2 Faculty of Natural Sciences and Mathematics of University of Tuzla, Tuzla, Bosnia and Herzegovina

References

[1] Adrovic F, Prokic M, Ninkovic M, Glišic R. Measurements of environmental Background radiation at location of coal-fired power plants. Radiation Protection Dosimetry. 2004;**112**(3):439-442, Oxford University

[2] Durrani S, Ilic R. Radon Measurements by Etched Track Detectors. World Scientific. Singapore; 1997

[3] UNSCEAR 1982. Sources and Effects of Ionizing Radiation, Annex C—Technologically Modified Exposure to Natural Radiation. New York: United Nations; 1982

[4] JP Elektroprivreda BiH d.d. Sarajevo: *Izvještaj o zaštiti okoline za 2009. godinu*

[5] AlphaGuard PQ2000 Radon Monitor. User Manual. Germany: Genitron Instruments; 2000

[6] Todorovic D, Popovic D, Nikolic J, Ajtic J. Radioactivity monitoring in ground level air in Belgrade urban area. Radiation Protection Dosimetry. 2010;**142**(2-4):308-313

Chapter 6

Radon Exposure and Human Health: What Happens in Volcanic Environments?

Diana Linhares, Patrícia Garcia and
Armindo Rodrigues

Additional information is available at the end of the chapter

http://dx.doi.org/10.5772/intechopen.71073

Abstract

Volcanic activity can cause hazardous effects to the environment and the health of the exposed persons such as an increased risk for the development of several cancers. In geothermal areas, volcanic gases such as radon are continuously vented from the main crater, from fumarolic fields or diffused through soil. The continued long-term exposure to radon can enhance the risk of lung cancer being considered the leading cause of lung cancer following tobacco smoking. The chronic exposure to volcanogenic radon requires the development of biomonitoring methods that will assist in the evaluation of the effects of exposure to this genotoxic element. The Human Biomonitoring with the use of exfoliated buccal cells is minimally invasive, and the endpoints of the buccal micronucleus cytome (BMCyt) assay are the biomarkers of effect most recently used to measure genetic damage for the exposure to genotoxic and cytotoxic xenobiotics. The BMCyt assay has been used in a number of occupational studies, and positive results were detected as a consequence of exposure to pesticides, metals, and industrial chemicals that are suspected to cause cancer. Regarding the chronic exposure to volcanic environments, many studies revealed a rise in the numbers of MN in buccal exfoliated cells, indicating an increased risk for cancer. This chapter aims to cover the main health hazards and biomonitoring methods for populations chronically exposed to volcanic environments, allowing an estimate of health risks and to implement risk management measures regarding the exposure to certain compounds.

Keywords: radon, volcanic environments, biomonitoring, BMCyt assay

1. Radon

1.1. Definition, physical and chemical properties

Radon was discovered by Fredrich Ernst Dorn in 1898. The name is derived from radium, as it was first detected as an emission from radium during radioactive decay. It is a naturally occurring radioactive gas of the noble gas family, is odorless and tasteless and, is produced by the decay of radium in the uranium decay chain. Radon is measured in terms of its activity (curies or becquerels), and these measurements provide information regarding how much a radioactive material decays every second (1 Ci = 37 billion Bq = 37 billion decays per second).

Radon also undergoes to radioactive decay, being divided in two parts (one part is called radiation, and the other part is called a daughter or progeny). In the air, the radioactive daughters (isotopes of polonium, bismuth, and lead) tend to attach to surfaces and to aerosol particles such as dust and cigarette smoke [1, 2, 3].

1.2. Sources in the environment

Radon is a radioactive compound, which rarely occurs naturally in the environment, being mostly derived from human activities; some of the potential sources of anthropogenic radiation include x-rays and other types of radiation used in medicine, radioactive waste generated by nuclear power stations and scientific research centers, and even electromagnetic radiation from television sets and microwave ovens.

On the other hand, radon isotopes are formed naturally through the radioactive decay of uranium or thorium, being present in rocks and soil. The decay products of ^{222}Rn, such as ^{218}Po and ^{214}Pb, can attach to particles in the air and be transported this way in the atmosphere being deposited on land or water by settling or by rain [4]. The human exposure to radon can occur by uranium mining activities (since uranium minerals emit radon gas), the occurrence of granitic bedrocks located underneath buildings [5], the use of building materials that contain variable uranium and radium concentrations, and exposure to volcanic activity.

1.3. Volcanic environments

During and after eruptions, volcanoes release hazardous gases such as sulfur dioxide (SO_2), sulfuric acid (H_2SO_4), hydrogen sulfide (H_2S), hydrogen chloride (HCl), hydrogen fluoride (HF), carbon dioxide (CO_2), and radon (Rn) [6]. These gases have both acute and chronic health effects: carbon dioxide and sulfur gases are the main gases responsible for acute mortality due to their asphyxiating and/or toxic properties; radon, due to its radioactivity, has important chronic effects, and thus, even at low doses, a long-term exposure may also pose a significant health risk.

According to Gasparini and Mantovani [7], the radon release, in volcanic systems, occurs mainly due to (a) the radon presence in rocks, minerals, pore fluids, and groundwater; (b) the release of magma fluids ascent to the surface, heating the nesting rocks; and (c) the increase in temperature that causes removal of the radon present in interstitial fluids and even into groundwater. Radon is considered a useful tracer of volcanic activity due to the ability to be transported from depth (by carrier gases such as CO_2) without being chemically altered [8]. However, volcanic environments are complex, and the nature of the driving force can change during gas ascent, depending on the physical-geological conditions in the environment that the gas encounters. Variations in temperature, pressure, mechanical stresses, chemical reactions, and mineral precipitation can change the gas-bearing properties of geological formations [9].

In a study developed in all regions of the Former Yugoslav Republic of Macedonia, it was shown that indoor radon concentration measurements in dwellings were significantly different between seasons, being the highest values observed in the winter and autumn [10]. Also, in the Vulsini Volcanic district in Northern Latium (Central Italy), Cinelli et al. [11] performed a series of soil gas radon measurements to estimate the radon radiation risk; the authors observed that soil gas radon concentration ranged between 7 and 176 kBq/m^3, revealing a large degree of variability particularly related to the different lithologies of the studied sites.

Considering that, in volcanically active environments, radon can be continuously vented from the volcano main crater, from fumarolic fields, or diffusely emitted through soil [12]; these environments present continuous health risks to the populations living in their vicinity, as it happens in the volcanic islands of the Azores archipelago.

The Azores archipelago is located in the North Atlantic Ocean, in the triple junction of the North American, African, and Eurasian plates [13], and is formed by nine islands of volcanic origin that represent the emerged part of the Azores Plateau. As a result of the Azores archipelago location on an active plate boundary, the islands are subjected to frequent seismic and volcanic activity, including eruptions and degassing processes. The São Miguel Island, located in the eastern group of the archipelago, has one of the most active and dangerous volcanoes of the archipelago, the Furnas volcano. Its activity is marked by the presence of fumarolic fields [14] and extensive soil diffuse degassing areas [15, 16] that surge the radon concentration increase [17]. In 2014, Silva et al. [18] surveyed the indoor radon concentration in some buildings from the villages of Furnas and Ribeira Quente (located in degassing areas in the vicinity of Furnas volcano); the annual radon concentration ranges between 23 and 6403 Bq/m^3. The level of indoor radon exceeds the limit defined by the regional legislation (150 Bq/m^3) (D.L.R No. 16/2009/A Art. 51) in 33% of the Furnas village buildings and in 21% of the ones of Ribeira Quente. Further up, Silva et al. [19] applied a series of regression models demonstrating that barometric pressure, soil water content, soil temperature, soil CO_2 flux, air temperature, relative air humidity, and wind speed are the statistical meaningful variables explaining between 15.8 and 73.6% of ^{222}Rn variations in Furnas volcano vicinity.

2. Human exposure to radon

Human exposure to natural ionizing radiation is largely due to radon. It is estimated that the worldwide radon contribution constitutes as much as 50% of the overall radiation dose, reaching values of about 1.15 mSv/year per capita [20].

Although the atmospheric concentration of radon is usually low, this gas tends to accumulate within confined environments with reduced or no air exchange, such as buildings, dwellings, tunnels, caves, and mines. During the radon decay process, alpha, beta, and gamma radioactive particles are released and can be inhaled and deposited on the bronchial epithelium of exposed individuals; considering that the alpha particle can disrupt the DNA structure within cells of the lining epithelia, and especially lung cells, exposure to this irradiation can contribute to an increased risk of lung cancer [17].

The effects of radon exposure in human health have been known since the sixteenth century, although radon was only discovered at the turn of the twentieth century. Still, it was only in the 1940s that a causal link between lung cancer in miners and radon exposure was established [21]. Taking in consideration that the studies on underground miners consistently demonstrated an increased risk of lung cancer caused by radon and its progeny, International Agency for Research on Cancer (IARC) classified radon as a human carcinogen in 1988 [22]. Later on in the 1970s, indoor radon accumulation in domestic buildings was first observed being considered to play a role in the epidemiology of lung cancer [23]. Considering that the awareness of the potential for hazardous exposure, due to accumulation of radon in indoor environments, has increased during the times, several programs were established to carefully monitor and control radon concentration in closed spaces, such as households located in areas where geogenic radon is highly likely to occur.

3. Biomonitoring

Traditionally, the term biomonitoring can be simply defined as a way to provide information about exposures to chemicals in living organisms. Exposure is commonly assessed by a spectrum of questionnaire data and ecological, environmental, or biological measurements. Biological measures of exposure allow the assessment of internal dose, by measuring the parent chemical or its metabolite or reaction product in the human blood, urine, milk, saliva, and adipose or other tissues [24]. Nowadays, and ever since Wild [25] proposed an environmental complement to the genome in determining risk of disease, termed the exposome (totality of exposures throughout the lifespan), researchers started to include in biomonitoring studies the cumulative measure of exposures to both chemical and nonchemical agents, such as diet, stress, and socio-behavioral factors. The exposomic approaches go a step beyond traditional biomonitoring, aiming to capture all exposures that potentially affect health and disease.

3.1. Human biomonitoring

When individuals or populations are exposed to a chemical, it may be absorbed and distributed among the bodily tissues, metabolized, and/or excreted (i.e., absorption, distribution, metabolism, and excretion (ADME)). For the assessment of the human exposure to a given xenobiotic, biologic measurements of the chemical can be made after the absorption step or during each of the subsequent steps of ADME [26]. This is called human biomonitoring (HBM), which is an approach for assessing human exposures to natural and synthetic compounds from the environment, occupation, and lifestyle; this scientifically developed methodology allows an evaluation of factors, such as cumulative exposure or genetic susceptibility, to a certain chemical compound allowing the extrapolation for adverse risks and health effects, such as cancer.

The data obtained in HBM studies integrate exposure from all routes (oral, inhalation, dermal, transplacental) and allow (1) establishing population reference ranges, (2) identifying unusual exposures for subpopulations, (3) evaluating temporal variability and trends within a population, (4) validating questions designed to estimate individual exposure, and (5) examining associations with health outcomes in epidemiologic studies. Therefore, HBM has tremendous utility, providing an efficient and cost-effective means of measuring human exposure to chemical substances, handing unequivocal evidence that both exposure and uptake have been taken place [27, 28].

3.2. Biomarkers

Biomarkers are commonly used as "agents" to measure the concentrations of chemical substances, their metabolites, or reaction products in human tissues or specimens, representing an integrative measurement of exposure to a given agent (i.e., the internal dose), that result from complex pathways of human exposure and also incorporate toxicokinetic information and individual characteristics such as a genetically based susceptibility [29].

In 1993, the WHO [30] provided three classifications of biomarkers—exposure, effect, and susceptibility—but with the rise of genomics and other advances in molecular biology, the biomarker classification has suffered some changes through time. In 2001, the Biomarker working group defined five classifications: (1) antecedent biomarkers, to identify the risk of developing an illness; (2) screening biomarkers, to screen for subclinical disease; (3) diagnostic biomarkers, to recognize overt disease; (4) staging biomarkers, to categorize disease severity; and (5) prognostic biomarkers, to predict future disease course. In this paradigm, biomarkers are viewed on a continuum between markers of exposure and markers of effect, with markers of susceptibility spanning the continuum.

The markers of effect are applied in the comparison of case and control, target and nontarget tissues, or dose and time to response, where a correlation between the biomarker and biological effect can be demonstrated; the markers of susceptibility are defined as the genetic factors that influence the body's sensitivity to a chemical but can also include biological factors, such as age, nutritional status, and lifestyle.

The continuous progress in molecular and analytical methods to measure biomarkers for linking exposure with health outcomes and new approaches in exposome research will enable a more comprehensive and integrated analysis across the biomarker continuum [31].

3.3. Micronucleus

Micronucleus (MN) is defined by Schmid [32] as a microscopically visible, round, or oval cytoplasmic chromatin mass next to the nucleus that is formed from acentric chromosomes, chromatid fragments, whole chromosomes, or chromatids. Micronucleus formation results from two basic phenomena in mitotic cells, which are chromosome breakage and the dysfunction of the mitotic apparatus.

This aberration is induced by genotoxic stress caused by a clastogen or an aneugen xenobiotic; the clastogen involves the induction of either chromosome fragments that lag behind the separating chromosomes or a chromatin bridge between chromosomes at the anaphase of mitosis; the aneugen causes a daughter cell to have an abnormal number of chromosomes by disrupting the whole chromosomes bound to the mitotic spindle at anaphase; by disrupting the spindle checkpoint, the chromatin is separated from the newly forming nucleus and forms an independent nucleus-like structure, the micronucleus [33].

Since this chromosomal aberration is a frequent and significant response to exposure to mutagenic agents, it has been extensively used to identify potential genotoxic exposures and also chromosomal instability [34].

3.4. Buccal micronucleus cytome (BMCyt) assay

The continuous chromosomal damage in fully differentiated cells (as in the buccal epithelial cells) implies that damage occurred in the basal cell layer during earlier nuclear divisions. Considering that buccal epithelial cells are in direct contact with inhaled and ingested substances and/or elements and more than 90% of all human cancers are of epithelial origin, it can be argued that buccal epithelial cells represent a preferred target site for early genotoxic events induced by carcinogenic agents [35]. Therefore, the MN assay using exfoliated buccal cells can be considered the most suitable biomonitoring approach for the detection of increased cancer risk in humans, when the exposure to genotoxic agents occurs via inhalation or ingestion.

The buccal micronucleus cytome (BMCyt) assay is a useful and minimally invasive method that allows screening for both genomic damage and cytotoxicity events [36]. This technique is minimally invasive [37] and is frequently used in biomonitoring studies of occupational and environmental exposure to carcinogenic substances [38–40].

This assay involves the collection of exfoliated buccal epithelial cells from inside the cheeks by scraping the oral mucosa, smearing onto glass slides, and staining by the Feulgen method; this methodology allows the examination of cells to determine the frequency of MN and other genomic damage markers such as karyorrhexis (nuclear disintegration associated with the loss of nuclear membrane integrity), karyolysis (nuclei completely depleted of DNA and therefore appear as Feulgen-negative ghost-like), and pyknosis (small shrunken, intensively stained nucleus); these anomalies are associated with both cytotoxicity (necrosis and keratinization) and genotoxicity (apoptosis) being considered as effective biomarkers for populations exposed to mutagenic and carcinogenic agents [41].

Factors such as the minimal invasiveness of cell collection, high reliability, and low cost of BMCyt have contributed to the worldwide success for epidemiological studies enabling the early detection of carcinogenic effects in the cell exposed to various carcinogenic agents [42].

4. Case studies

It has been demonstrated that exposure to radon can induce lung tumors and other cancers in experimental animals [43]. Also, a radon dose-related carcinogenicity has been shown through epidemiologic studies of miners and case-control studies in the general population [44–47].

In Europe, Darby et al. [48] in a collaborative study that included 13 case-control studies in 9 European countries, each of which registered over 150 people with lung cancer and 150 or more controls and which included data about radon levels over 15 years, found that the risk of lung cancer (after stratification for study, age, gender, region of residence, and smoking) increased by 8% per 100 Bq/m^3 increase in radon concentration. In the United States, Krewski et al. [49] evaluated the risk associated with prolonged residential radon exposure, combining data from seven large-scale case-control studies (4081 cases and 5281 controls), and these authors also found that the risk of lung cancer due to exposure to residential radon was increased by 10% per 100 Bq/m^3.

For the biomonitoring of the human populations, the application of easy and noninvasive techniques is necessary; for the last 10–15 years, there has been an increase in the use of cytogenetic monitoring techniques, such as the buccal micronucleus cytome (BMCyt) over the use of MN in lymphocytes. Nevertheless, there are only a few studies, developed in the Azores, that apply this assay for the risk assessment of the chronic exposure to radon of volcanic origin.

Considering the geologic context of the archipelago, some studies have been developed to biomonitor the health effects of the chronic exposure the compounds released by the volcanic activity. In the study of Amaral et al. [50], it was observed that the population living in the vicinity (inside the crater) of Furnas volcano presented higher rates of cancer of the lip, oral cavity, and pharynx, in both sexes, and of female breast cancer. The relative risk analysis revealed a higher risk for lip, oral cavity, pharynx, and breast cancer in this population compared to a population inhabiting an island without historical records of volcanic activity. According to these authors, these higher cancer rates could be partially explained by the chronic exposure to environmental factors resulting from volcanic activity. More recently, Rodrigues et al. [51] in a study carried with inhabitants of Furnas volcano demonstrated that the chronic exposure to a volcanically active environment results in genotoxic and cytotoxic effects in human oral epithelial cells; this research group used the buccal micronucleus assay and observed that the frequencies of micronucleus and other nuclear anomalies were significantly higher in the Furnas inhabitants when compared to the inhabitants of an area without volcanic activity. Also, the inhabitants of Furnas village presented a 2.4-fold higher risk of having a higher frequency of micronucleated cells.

More recently, Linhares et al. [52] took a step forward in this investigation and evaluated the DNA damage in the buccal epithelial cells of individuals chronically exposed to indoor radon in Ribeira Quente inhabitants, a village within 7 km (in the south flank) of the Furnas volcano. The authors shown that indoor radon concentration correlated positively with the frequency of micronucleated cells and that the exposure to indoor radon is a risk factor for the occurrence of micronucleated cells in the inhabitants of the hydrothermal area.

The results obtained in these studies demonstrate a significant association between chronic exposure to indoor and the occurrence of DNA damage evidencing the usefulness of biological surveillance to assess mutations involved in pre-carcinogenesis, reinforcing the need for further studies with human populations. Likewise, these studies set the basis for the use of the BMCyt assay in volcanic areas worldwide to assess the effects of the chronic exposure to genotoxic substances and/or elements.

5. Risk management

It has been reported that more than 60% of the ionizing radiation a person receives every year can be caused by natural sources of radiation and more than 50% of this radiation can be due to radon and the products of its disintegration. Hence, maintaining radon safety is a critical challenge, and this has been actively discussed throughout the last few decades [53]. To assure a public health risk assessment, factors such as the source, the pathway, and the receptor must be considered.

The indoor radon concentration is associated with factors such as the building structure (such as construction material, pavement, ventilation, and the presence of basements or air boxes), as well as the ventilation measures taken. These factors are particularly relevant in households built on volcanic hydrothermal areas, where soil diffuse degassing can be very intense. As for the consequences of human exposure to indoor radon, these will depend of factors such as the concentration of radon indoors, the duration of exposure, the general health status of the individual, and the synergistic effect with smoking.

In order to manage the risk, primary strategies such as education and prevention activities must be established to prevent/diminish the entrance of radon into the buildings [54]. However, this may not be easily achieved, since the indoor radon varies considerably by region and locality and is greatly affected by the household structure as well as soil and atmospheric conditions. Consequently, often the development of secondary strategies to remediate the effects of indoor radon is required, such as taking measures in the construction of the household by selecting an adequate foundation type, location, building materials used, entry points for soil gas, and building ventilation systems.

At an individual level, it is necessary to ensure an adequate past medical history or family medical history, since individuals who have chronic respiratory diseases such as asthma, emphysema, and fibrosis may be more susceptible to the respiratory effects of radon and radon progeny. Due to the reduced expiration efficiency and the increased residual air volume of these individuals, radon and its progeny would be resident in the lungs for longer periods of time, increasing the risk of damage to the lung tissue.

The above-discussed epidemiologic studies clearly evidence the carcinogenicity of radon in volcanically active environments. However, radon control efforts appear to be held up, and

http://dx.doi.org/10.5772/intechopen.71073

the cost-effectiveness of current strategies and interventions is questionable; therefore, the continuous biomonitoring of the populations exposed to radon in volcanic areas is necessary to further implement risk management measures.

Author details

Diana Linhares[1,2*], Patrícia Garcia[1,3] and Armindo Rodrigues[1,2]

*Address all correspondence to: diana.ps.linhares@uac.pt

1 Faculty of Sciences and Technology, University of the Azores, Ponta Delgada, Portugal

2 IVAR, Institute of Volcanology and Risks Assessment, University of the Azores, Ponta Delgada, Portugal

3 cE3c, Centre for Ecology, Evolution and Environmental Changes, and Azorean Biodiversity Group, University of the Azores, Ponta Delgada, Portugal

References

[1] Edling C (1985) Radon daughter exposure and lung cancer. British Journal of Industrial Medicine, **42**:721-722

[2] Biermann AH, and Sawyer SR (1995) Attachment of radon progeny to cigarette-smoke aerosols. Lawrence Livermore Natl. Lab., CA (United States). Technical Report UCRL-CR-120647

[3] Ojo TJ, and Ajayi IR (2014) A review on environmental radon and its potential health risk on humans. Journal of Environmental Science, Toxicolology and Food Technology, **8**, 1-8

[4] Agency for Toxic Substances and Disease Registry (ATSDR) (2012) Public Health Statement for Radon. Atlanta, GA: U.S. Department of Health and Human Services, Public Health Service

[5] Denton GRW and Namazi S (2013) Indoor Radon Levels and Lung Cancer Incidence on Guam. Procedia Environmental Sciences, **18**:157-166

[6] Durand M, Florkowski C, George P, Walmsley T, Weinstein P, Cole J (2004) Elevated trace element output in urine following acute volcanic gas exposure. Journal of Volcanology and Geothermal Research, **134**:139-48

[7] Gasparini P and Mantovani MSM (1978) Radon anomalies and volcanic eruptions. Journal of Volcanology and Geothermal Research, **3**, 325-341

[8] Barbosa SM, Donner RV, Steinitz G (2015) Radon applications in geosciences–Progress & Perspectives. The European Physical Journal Special Topics, **224**:597-603

[9] Padilla GD, Hernández P, Padrón E, Barrancos J, Pérez NM, Melián G et al. (2013) Soil gas radon emissions and volcanic activity at El Hierro (Canary Islands): The 2011-2012 submarine eruption. Geochemistry, Geophysics, Geosystems, **14**:432-447

[10] Stojanovska Z, Januseski J, Bossew P, Zunic ZS, Tollefsen T, Ristova M (2011) Seasonal indoor radon concentration in FYR of Macedonia. Radiation Measurements, **46**:602-610

[11] Cinelli G, Tositti L, Capaccioni B, Brattich E, Mostacci D (2015) Soil gas radon assessment and development of a radon risk map in Bolsena, Central Italy. Environmental Geochemistry and Health, **37**(2):305-319

[12] Hansell A, Horwell C, Oppenheimer C (2006) The health hazards of volcanoes and geothermal areas. Occupational and Environmental Medicine, **63**:149-156

[13] Searle R (1980) Tectonic pattern of the Azores spreading centre and triple junction. Earth and Planetary Science Letters, **51**:415-434

[14] Ferreira T, Gaspar JL, Viveiros F, Marcos M, Faria C, Sousa F (2005) Monitoring of fumarole discharge and CO2 soil degassing in the Azores: contribution to volcanic surveillance and public health risk assessment. Annals of Geophysics, **48**:4-5

[15] Viveiros F, Cardelllini C, Ferreira T, Caliro S, Chiodini G, Silva C (2010) Soil CO2 emissions at Furnas Volcano, São Miguel Island, Azores archipelago: Volcano monitoring perspectives, geomorphologic studies, and land use planning application. Journal of Geophysical Research, **115**, B12208, 1-17

[16] Viveiros F, Cardellini C, Ferreira T, Silva C (2012) Contribution of CO2 emited to the atmosphere by diffuse degassing from volcanoes: The Furnas Volcano case study. International Journal of Global Warming, **4**(3-4): 287-304

[17] Appleton JD (2013) Radon in air and water. In: Essentials of medical geology. O. Selinus et al. (Eds), Dordrecht, Netherlands, pp. 239-277. Springer Netherlands

[18] Silva C, Ferreira T, Viveiros F, Allard P (2014) Indoor radon (222Rn) monitoring in some buildings at Furnas and Ribeira Quente villages (Furnas Volcano, Azores): public health risk assessment. Comunicações Geológicas, **101**, Especial II, 927-931

[19] Silva C, Ferreira T, Viveiros F, Allard P (2015) Soil radon (222Rn) monitoring at Furnas Volcano (São Miguel, Azores): applications and challenges. The European Physical Journal Special Topics, **224**(4): 659-686

[20] UNSCEAR (United Nations Scientific Committee on the effects of Atomic Radiation). (2008). Sources and effects of ionizing radiation. Report to General Assembly, Annex B, United Nations, New York

[21] Hueper WC (1942) Occupational Tumors and Allied Diseases. Springfield, Ill. Charles C Thomas

[22] International Agency for Research on Cancer Monographs on the Evaluation of Carcinogenic Risks to Humans (IARC) (1988) Man-Made Mineral Fibres and Radon. Lyon, France: IARC Press; **43**:33-171

[23] Axelson, O (1984) Room for a role for radon in lung cancer causation? Medical Hypotheses, **13**(1):51-61

[24] Needham LL, Barr DB, Calafat AM (2005) Characterizing children's exposures: beyond NHANES. Neurotoxicology, **26**:547-553

[25] Wild CP (2005) Complementing the genome with an "exposome": the outstanding challenge of environmental exposure measurement in molecular epidemiology. Cancer Epidemiology, Biomarkers and Prevention, **14**:1847-1850

[26] LaKind JS, Sobus JR, Goodman M, Barr DB, Furst P, Albertini RJ, et al. (2014) A proposal for assessing study quality: biomonitoring, environmental epidemiology, and short-lived chemicals (BEES-C) instrument. Environment International, **73**:195-207

[27] Sexton K, Needham L, Pirkle J (2004) Human Biomonitoring of Environmental Chemicals. American Scientist, **92**, 38

[28] Angerer J, Ewers U, Wilhelm M (2007) Human biomonitoring: state of the art. International Journal of Hygiene and Environmental Health, **210**(3-4): 201-228

[29] Ladeira C and Viegas S (2016) Human Biomonitoring – An overview on biomarkers and their application in Occupational and Environmental Health. Biomonitoring, **3**:15-24

[30] World Health Organization (WHO). 1993. Biomarkers - Biomarkers and risk assessment: Concepts and principles, Environmental Health, Criteria No.155(93), Geneva

[31] Dennis KK, Marder ME, Balshaw DM, Cui Y, Lynes MA, Patti GJ, et al. (2017) Biomonitoring in the era of the exposome. Environmental Health Perspectives, **125**(4), 502-510

[32] Schmid W (1975) The micronucleus test. Mutation Research, **31**(1):9-15

[33] Utani K, Kohno Y, Okamoto A, Shimizu N (2010) Emergence of Micronuclei and Their Effects on the Fate of Cells under Replication Stress. PLoS ONE, **5**(4): e10089. https://doi.org/10.1371/journal.pone.0010089

[34] Tolbert PE, Shy CM, Allen JW (1992) Micronuclei and other anomalies in buccal smears: method development. Mutation Research, **271**:69-77

[35] Holland N, Bolognesi C, Volders MK, Bonassi S, Zeiger E, Knasmueller S (2008) The micronucleus assay in human buccal cells as a tool for Biomonitoring DNA damage: The HUMN project perspective on current status and knowledge gaps. Mutation research, **659**:93-08

[36] Thomas P, Holland N, Bolognesi C, Kirsch-Volders M, Bonassi E, Zeiger S, et al. (2009) Buccal micronucleus cytome assay. Nature Protocols, **4**:825-837

[37] Bolognesi C, Knasmueller S, Nersesyan P, Thomas P, Fenech M (2013) The HUMNxl scoring criteria for different cell types and nuclear anomalies in the Buccal Micronucleus Cytome Assay: an update and expanded photogallery. Mutation Research-Reviews in Mutation Research, **753**(2):100-13

[38] Çelik A, Yildirim S, Ekinci SY, Taşdelen B (2013) Bio-monitoring for the genotoxic assessment in road construction workers as determined by the buccal micronucleus cytome assay. Ecotoxicology and Environment Safety, **92**:265-270

[39] Hallare AV, Gervasio MK, Gervasio PL, Acacio-Claro PJ (2009) Monitoring genotoxicity among gasoline station attendants and traffic enforcers in the City of Manila using the micronucleus assay with exfoliated epithelial cells. Environmental Monitoring and Assessment, **156**:331-341

[40] Garcia PV, Linhares D, Amaral AFS, Rodrigues AS (2012) Exposure of thermoelectric power-plant workers to volatile organic compounds from fuel oil: Genotoxic and cytotoxic effects in buccal epithelial cells. Mutation Research, **747**:197-201

[41] Fenech M (2007Cytokinesis-block micronucleus cytome assay. Nature Protocols, **2**: 1084-1104

[42] Bonassi S, Ugolini D, Kirsch-Volders M, Strömberg U, Vermeulen R, et al. (2005) Human population studies with cytogenetic biomarkers: review of the literature and future prospectives. Environmental and Molecular Mutagenesis, **45**:258-270

[43] Monchaux G, Morlier J, Morin M, Chameaud J, Lafuma J, Masse R (1994) Carcinogenic and Cocarcinogenic Effects of Radon and Radon Daughters in Rats. Environmental Health Perspectives, **102**(1):64-73

[44] Schoenberg JB, Klotz JB, Wilcox HB, Nicholls GP, Gil-del-Real MT, Stemhagen A, et al. (1990) Case–control study of residential radon and lung cancer among New Jersey women. Cancer Research, **50**(20):6520-6524

[45] Lubin JH, Boice Jr JD, Edling C, Hornung RW, Howe GR, Kunz E, Kusiak RA, Morrison HI, Radford EP, Samet JM et al. (1995) Lung cancer in radon-exposed miners and estimation of risk from indoor exposure. Journal of the National Cancer Institute, **87**(11):817-27

[46] Duport P (2002) Is the radon risk overestimated? Neglected doses in the estimation of the risk of lung cancer in uranium underground miners. Radiation Protection Dosimetry, **98**(3):329-338

[47] Wichmann HE, Rosario AS, Heid IM, Kreuzer M, Heinrich J, Kreienbrock L (2005) Increased lung cancer risk due to residential radon in a pooled and extended analysis of studies in Germany. Health Physics, **88**(1):71-79

[48] Darby S, Hill D, Auvinen A, Barros-Dios JM, Baysson H, Bochicchio F, et al. (2005) Radon in homes and risk of lung cancer: collaborative analysis of individual data from 13 European case-control studies. British Medical Journal, **330**(7485): 223-226

[49] Krewski D, Lubin JH, Zielinski JM, Alavanja M, Catalan VS, William Field R, et al. (2006) A combined analysis of North American case-control studies of residential radon and lung cancer. Journal of Toxicology and Environmental Health, Part A, **69**(7-8), 533-597

[50] Amaral A, Rodrigues V, Oliveira J, Pinto C, Carneiro V, Sanbento R, et al. (2006) Chronic exposure to volcanic environments and cancer incidence in the Azores, Portugal. Science of the Total Environment, **367**:123-128

[51] Rodrigues A, Arruda M, Garcia P (2012) Evidence of DNA damage in humans inhabiting a volcanically active environment: A useful tool for biomonitoring. Environment International, **49**:51-56

[52] Linhares D, Garcia P, Silva C, Barrosos J, Kazachkova N, Pereira R, Lima M, Camarinho R, Ferreira T, Rodrigues A (2016) DNA damage in oral epithelial cells of individuals chronically exposed to indoor radon (222Rn) in a hydrothermal area. Environmental Geochemistry and Health. DOI: 10.1007/s10653-016-9893-2

[53] Bissett RJ and McLaughlin JR (2010) Radon. Chronic Diseases in Canada, **29**:38-50

[54] Larsson LS (2014) Risk-Reduction Strategies to Expand Radon Care Planning with Vulnerable Groups. Public Health Nursing (Boston, Mass.), **31**(6), 526-536

Chapter 7

Residential Radon Exposure and Lung Cancer Risk in Kazakhstan

Rakhmetkazhy Bersimbaev and Olga Bulgakova

Additional information is available at the end of the chapter

http://dx.doi.org/10.5772/intechopen.71135

Abstract

Many published data have demonstrated the potential of radon exposure to induce biological damage. All these studies suggest that exposure to radon and its progeny can represent a significant public health risk. Radon exposure is considered as the second cause of lung cancer and it is the first in never smokers. Many countries have depicted residential radon exposure maps in order to characterize those areas with the highest indoor radon concentrations. There are areas in Kazakhstan with a number of factors leading to natural radioactivity. The genotoxic effects of radon on population of Kazakhstan are poorly understood. These studies conducted in compliance with all requirements of genotoxic studies will provide an extensive and reliable data for a detailed radon zoning of Kazakhstan. Our review attempts to integrate data about the association between residential radon levels and morbidity of population in Kazakhstan. In addition, we tried to cover all points of the cytogenetic and molecular changes induced by radon exposure.

Keywords: radon, uranium, lung cancer, p53, miRNA

1. Introduction

The radioactive contamination is a significant factor affecting the environment and human health. Radon and its decay products are the major contributors to human exposure from natural radiation sources. WHO has identified the chronic residential exposure to radon and its decay products as the second cause of lung cancer after tobacco consumption and the most common cause of lung cancer in non-smokers. Radon is a chemically inert radioactive gas, occurring naturally as an indirect decay product of uranium. Uranium has been actively mined and milled in the Republic of Kazakhstan [1]. Uranium ore mining and processing were initiated in Kazakhstan shortly after Second World War, and lasted for almost half a century. It has been estimated that during the Soviet period, about 30–40% of the uranium was extracted from the Asian region [2, 3].

Accordingly, the population of Kazakhstan might be exposed to a variety of hazardous materials including radon. Almost the whole eastern and northern part of Kazakhstan is potentially radon-affected area. The analysis of the cancer rate on the territory of Kazakhstan showed that there is a significant relationship between the areas with high level of morbidity and radon-affected areas. It is worth emphasizing that lung cancer is ahead of other cancer diseases in some areas of Kazakhstan that are radon-affected territories.

Lung cancer in the structure of oncology occupies the second rank position, and its share at the moment was 11.4%. This is the most widely distributed form of cancer leading since 1985 in Republic. The highest morbidity of lung cancer is observed in the North Kazakhstan, Pavlodar, Akmola, and East Kazakhstan regions. Almost the whole of Kazakhstan to the east of the Kustanai-Shymkent line (North Kazakhstan, Pavlodar, Akmola, East Kazakhstan, Kustanai and Karaganda regions) is potentially radon-affected areas. Thus, it becomes evident that one of the measures to reduce the incidence of lung cancer is to identify areas with high level of radon and reduce the impact of radon on the population.

When radon is inhaled, α-particles could affect lung tissue and may cause lung cancer via DNA and oxidative damages [4]. However the molecular effect of radon on lung cancer is not understood.

In addition to the mutations that arise in oncogenes or tumor-suppressor genes that are necessary for tumor transformation, epigenetic changes can play a key role in the pathogenesis of cancer. At the epigenetic level, specific alterations including alterations in DNA methylation, histone modifications leading to generation of γ-H2AX species (indicative of double-strand DNA breaks) and miRNAs expression patterns have been associated with radon exposure. Nevertheless, the mechanistic details underlying the molecular pathways by which genetic and epigenetic mechanisms are associated with lung cancer remain unclear. This review attempts to integrate data about the association between residential radon levels and morbidity of population in Kazakhstan living in these regions. In addition, we tried to cover all the points of the cytogenetic and molecular changes induced by radon exposure.

2. Uranium pollution in Kazakhstan

Uranium (U) is a radioactive chemical element of group III of the periodic system with atomic number 92 that belongs to the family of actinides. Uranium is readily soluble in water, easily reacts with the basic elements of organic compounds. In nature, uranium is found as uranium-238, uranium-235, and a very small amount of uranium-234.

The Republic of Kazakhstan is rich for uranium deposits and second in the world for mining resource, which contributes to the pollution of the environment with uranium tailings. Total deposits are estimated to be 2 million tons, representing 25% of uranium deposits in the world. The mining and milling of uranium ore have caused contamination of the environment through a number of activities, especially the open-pit mining process, transportation to and from milling sites, the milling and processing of ore, and open-air storage of radioactive and non-radioactive mining wastes. The main concern in Kazakhstan is that approximately 85% of the urban population lives in territories where environmental pollution is known to exceeded permissible standards [5, 6] together with vastly unknown contamination problems.

The largest region enriched with the uranium deposits is located within the North Kazakhstan containing approximately 16.4% of the uranium resources of Republic [7]. Besides, there are also large storage facilities for the radioactive waste in the North and West Kazakhstan. The total area exposed to radioactive waste from the uranium industry enterprises is estimated to approximately 100,000 hectares with a total activity of 250,000 Curie. In mining, the uranium and its decay products buried deep in the earth are brought to the surface, and the rock containing them is crushed into fine sand. After the uranium is chemically removed, the sand is stored in huge reservoirs [8]. Conservation and liquidation of uranium deposits in northern Kazakhstan was completed in 2008 in accordance with the State program "Preservation of non-uranium mining enterprises and liquidation of the consequences of uranium deposits for 2001–2010" [9].

The decay chain of uranium is commonly called the "uranium series" or "radium series" and includes protactinium, radium, thorium, and radon.

The high levels of radon are observed in the North and East areas of Kazakhstan because of the natural radiation sources and the long-term and large-scale mining of uranium.

3. Characteristic of radon and its decay product

Radon (Rn) is a radioactive colorless and odorless inert gas. In nature, radon occurs in two forms: radon-222 (^{222}Rn), formed by the decay products of uranium-238; and radon-220 is the product of the decay of thorium-232. However, most of the radioactivity in the atmosphere at sea level is attributable to ^{222}Rn. For that reason, the term "radon" identifies mainly the ^{222}Rn.

The main sources of indoor radon are soil, building materials, water sources, natural gas, and outdoor air. There is a large variation of indoor air concentrations of this gas, mainly depending on the geology of the area and on factors affecting the differential pressure between inside and outside of the buildings, such as ventilation rates, heating systems, and meteorological conditions. The rocks with high radioactivity and tectonic faults with the high emanation can lead to a significant increase of radon concentration indoor and its contribution becomes dominant in the collective dose of the population.

4. Indoor radon data in Kazakhstan

According to the experts, the contribution of natural sources in the average annual radiation dose of the Kazakhstan population currently stands at 80%, including 50% from radon. The level of average annual radiation dose due to radon for the population of Kazakhstan is considered to be 1.5 times higher than the world average [10].

As showed in Summary Report "Legacy of uranium mining activities in central Asia—contamination, impact and risks", the highest dose contribution to humans was derived from indoor radon concentrations in houses and dwellings in the vicinity of the Kurday site. The values of indoor and outdoor radon concentrations in Kurday are considerably higher than global average corresponding values. The indoor radon concentrations (70–330 Bq/m^3) were found to be higher than the outdoor radon concentrations (30–90 Bq/m^3) [11].

In some settlements in 70% of the buildings, radon concentration exceeds the maximum permissible concentration (200 Bq/m^3). There are official data showing that the concentration of radon in the soil in some areas of Kazakhstan reaches 300,000 Bq/m^3, and the concentration of indoor reaches 6000–12,000 Bq/m^3, which exceeds the maximum permissible concentration by 60 times [12]. The radon-affected areas in Kazakhstan are shown in **Figure 1**.

Stegnar et al. [2] measured indoor ^{222}Rn concentrations in 27 selected houses and public buildings in Ust-Kamenogorsk city in East Kazakhstan area. According to their data, the values of radon ranged from 22 to 2100 Bq/m^3, the average concentration of radon was 230 Bq/m^3. The corresponding annual effective doses ranged from 0.5 to 20 mSv, the average dose being

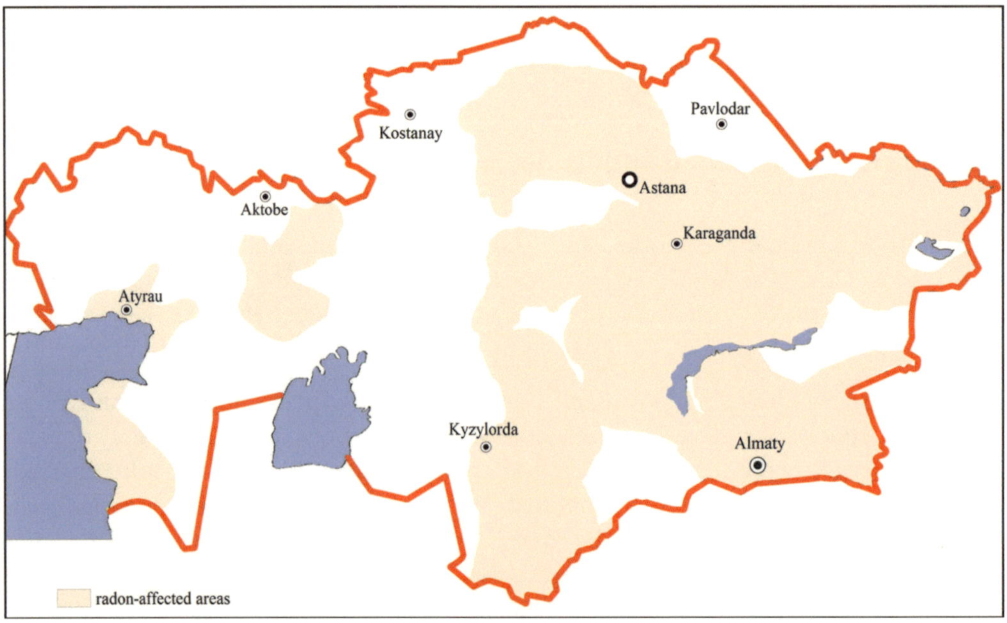

Figure 1. Radon-affected areas of Kazakhstan.

5 mSv per year. The studies conducted in Altai (East Kazakhstan) revealed radon levels from 200 to 8000 Bq/m^3 in 15 of the 18 settlements [13].

The territory of North Kazakhstan is characterized by the areas with the high levels of radiation, arising both from natural radiation sources, as well as by long-term and the large-scale activities of uranium mines and uranium-processing companies [11].

The Akmola region (North Kazakhstan) is one of the largest world's uranium ore provinces. It contains more than 30 uranium deposits [12]. During 50 years, open and underground mining of uranium ore was conducted in the North Kazakhstan that led to the accumulation of the significant amount of radioactive waste with the high risk of radioactive contamination to the environment and toxic effects on the human health. All these factors contribute to the formation of elevated concentrations of radon in this region. The measurement of the radon activity in indoor air was carried out in 2010 in the territory of three districts of Akmola region. As a result of this work, 47 settlements regions were revealed, including the Astana and Kokshetau cities, 35 settlements (76.1%) which were characterized by excess of standard values (200 Bq/m^3) radon activity [12].

In 2010, 814 measurements of radon in indoor air of the residential and public buildings were carried out in Astana in order to implement the preventive health surveillance. Exceeding levels of radon has been established in eight newly built homes (207–1150 Bq/m^3). It also revealed that excess radon activity in the Balkashino village school (1022 Bq/m^3) and in the preschool of the village "Shantobe" (789 Bq/m^3). It revealed maximum values of ^{222}Rn activity in the settlements: Vasilkivka—866 Bq/m^3, Granite—611 Bq/m^3, and Ondiris—419 Bq/m^3. In the village of Aksu, there were 42 surveyed areas of which 22 cases were detected exceeding the permissible radon activity. Activity of radon in homes ranged from 8 to 858 Bq/m^3, in schools—153–2162 Bq/m^3, and in the basement—130–5870 Bq/m^3. In Saumalkol village, the value of radon activity was in the range from 590 to 3977 Bq/m^3. In some kinder gardens of Aryk-Balyk village, radon concentrations range from 510 to 4500 Bq/m^3, that is in 22.5 higher than the allowable level of 200 Bq/m^3. The most famous case of a high level of radon in mines and in a residential area in Kazakhstan is mine "Akchatau". The major sources of ^{222}Rn concentration, in the opinion of Soroka and Molchanov, in dwellings in Akchatau village was condition radon flux from the surface of the granite massive (especially in a zone of geological rupture) and radioactivity of local construction materials (granite and adobe) including sand from mill plant tailings. The high level was of radon in 44% of the residential buildings in Akchatau village with a peak concentration at 37,650 Bq/m^3 [14]. It was under a comprehensive survey of radon in mines and homes, as well as the health of miners and residents. As a result, it has been found that the incidences of respiratory diseases, nervous system, and cardiovascular system exceed the average for the region is 2.9 times. It is most likely indicative of radon exposure on the health of the village.

A large group of the population is exposed to radon in the workplace. It is interesting that the workers employed in the mining and processing of non-ferrous metals most affected by radon, but not the workers of uranium mines. This is due to the fact that, there are high concentrations of naturally radioactive dust in the air. In some cases, exposure doses for lungs of underground miners exceeded 5 Sv y^{-1}, while the allowable level is 0.015 Sv y^{-1} [15].

In Almaty city, during the period from 1990 to 1993, preliminary measurements of level of radon in public buildings and kindergartens were carried out. Several places with level of radon between 380 and 532 Bq/m^3 were identified [16]. In some other areas of the North Kazakhstan region (Akkol, Yenbekshilder, and Sandyktau villages), 112 settlements were examined. More than 60% of the settlements were identified with exceeding activity of radon in indoor air and the drinking water sources in 32% of villages.

5. Radon concentrations in drinking water in Kazakhstan

The source of drinking water in most parts of Kazakhstan is groundwater and in many settlements there are no alternative sources of drinking water. However, often underground water is saturated with uranium, radium, radon, the concentrations of which exceed the permissible level. Therefore, a survey of water sources and water quality control are extremely relevant for preventing radiation risk of the population.

The active study of the concentrations of the natural radioactive elements in natural waters in Kazakhstan began in the late 1940s of last century due to an increase in geological prospecting for radioactive materials. From 1960 to 1992, the entire territory of Kazakhstan was covered by the hydrogeological surveys for water supply of the settlements. To carry out these activities in large volumes, the strictly regulated selections of water samples were conducted for determination of uranium, radium, and radon. The amount of available information in the archives and reports of specialized organizations includes information on approximately 30,000 water sources [5]. It has been shown that in the drinking water of some settlements, specific activity for ^{238}U reached 96 Bq/l, activity for ^{226}Ra 45 Bq/kg, and activity for ^{222}Rn 5100 Bq/kg [17].

In Kazakhstan, there are numerous sources of radon waters, which are determined by the presence of highly radioactive rocks enriched in uranium. Basically, water sources with a high content of radon (more than 100 Bq/l) are located in the submontane and mountain regions in the south, south-east, and north of Kazakhstan. These are uranium provinces Kendyktas-Chuili-Betbakdalin (Zhambyl, Almaty regions), North Kazakhstan (Akmola region) and the North-Pribalkhash rare-metal area (Karaganda region). In these regions, especially in villages near the mines, radiological services have identified the most numerous cases of high radon level in the air of residential and industrial premises.

The radiological evaluations of drinking water were carried out according the norms established by "Sanitary requirements for radiation safety" representing the main technical document regulating norms of ionizing radiation in Kazakhstan. According to this document, the contents of the natural and artificial radionuclides in drinking water, creating an effective dose less than 0.1 mSv/year does not require measurements to reduce the radioactivity of drinking water. Over the last years in Akmola region (North Kazakhstan), 1271 drinking water samples were investigated, from which 550 samples did not meet the standards for total alpha activity (43%). On average, the radiation dose of the population received drinking water is 0.21 mSv/year in this region [12]. Thus, studies on groundwater sources in North Kazakhstan have shown the presence of radon in drinking water sources from 8 to 300 Bq/l. A total of 32 surveyed drinking sources in 15 cases (47%) had high levels of radon in drinking water [12].

The studies of groundwater in Ust-Kamenogorsk (East Kazakhstan), conducted in 2005, showed the average level of radon (^{222}Rn) 60 Bq/l [12]. The concentration of radon in drinking groundwater may be associated with the presence of water in its long-lived daughter product ^{210}Pb, which is one of the most radiotoxic beta-active natural radionuclides.

6. Health effect of uranium on the population in Kazakhstan

The analysis of medical statistics of North Kazakhstan showed that the level of overall mortality and cancer among adults and children is one of the highest in the region [18], although it has not revealed a direct relationship between these indicators and the radiation factor as a result of uranium mining and processing enterprises.

The production of uranium involves a large contingent of workers whose work takes place in the specific conditions exposing to radiation and chemical factors present in the ore mined. Klodzinskiy et al. [19] have shown a deterioration of health of uranium miners in North Kazakhstan. The results of epidemiological and medical studies found the high frequency of respiratory diseases (61%) in the cohort of uranium miners in North Kazakhstan. According to Ref. [20], the most common somatic diseases among uranium workers of Stepnogorsk city were hypertension, chronic obstructive pulmonary disease, and chronic gastritis. Moreover dysfunction of all parts of the immune system was observed in the studied groups of uranium workers [21]. Comprehensive clinical assessment of health status of children and adults living near Stepnogorsk uranium-processing enterprises showed a high risk of developing chronic diseases in internal organs due to long-term toxic effects of radiation. The study of reproductive health of women living in areas adjacent to the uranium mines showed a low percentage of healthy women (13%) [22]. Among children in Stepnogorsk city, the urinary tract infections and chronic bronchitis amounted to a high percentage of the total somatic pathology [23].

In our previous study [24, 25], uranium mine workers in the Stepnogorsk mining-milling complex in North Kazakhstan were investigated for the expression of chromosome aberrations and for genetic factors that can modify the exposure-related expression of chromosome damage. The study has demonstrated an increased frequency of chromosomal aberrations in uranium miners compared to the matched control subjects, which were not exposed to uranium compounds or to any other known chemical or physical mutagens. The study of chromosomal aberrations in the workers of uranium mining showed a high frequency of acentric chromosomes, dicentric and centric ring chromosomes in dose/ dependent manner.

The uranium workers were classified into three groups according to their duration of exposure: group I: 1–10 years, II: 11–20 years, and III: 21–25 years. The received data show that all three groups of workers had higher frequencies of chromosomal aberrations than the control group. In the first group of workers, the frequency of dicentrics was higher (1.95 ± 0.15) than that of the matched control group (0.50 ± 0.09) ($0.05 < p < 0.01$). In the second group of uranium workers, the frequency of dicentrics and the total chromosomal aberrations were significantly higher (2.33 ± 0.17) than the control group (0.55 ± 0.08). In the third group of workers, the frequency of chromosomal aberrations was also significantly higher (2.68 ± 0.26) than the control group (0.36 ± 0.10). The frequency of chromosomal aberrations observed in exposed group III was also higher compared to the first and second exposed groups ($0.05 < p < 0.01$).

Furthermore, a significant increase in the frequency of chromosomal aberrations in the workers who were heavy smokers was observed compared with those who were moderate smokers or non-smokers. Uranium-exposed workers who had inherited the null GSTM1 and/or GSTT1 genotypes had a significant increase in the frequency of chromosome aberrations compared with those who had intact GSTM1 and GSTT1 genes for different group of workers [26]. We have shown that the uranium mining conditions in North Kazakhstan can cause a long-term health risk among the workers. In uranium miners, the various forms of malignant tumors of the lung, liver, and stomach were observed [27].

Lung cancer is the leading cause of cancer death and the second most common cancer among population in the Kazakhstan (11.4%). The incidence of lung cancer in men is higher than in women and accounts for 20% of the total number of cancer diseases. Over past 30 years in East Kazakhstan, there was a sharp increase in cancer rates. This growth is particularly noticeable in the Ust-Kamenogorsk city. The increase in total cancer morbidity indicator is mainly due to an increase in the incidence of lung cancer, breast cancer, and skin cancer [28].

As mentioned above, the villages where there is a high radon activity, marked by the territory of Kazakhstan, are characterized by the highest levels of radon in East and North Kazakhstan (67% of settlements in excess). Interestingly, in the Eastern and Northern part of the country, there is an increase (up to 1.5 times or more) in cancer rates, which makes it possible to assume a correlation between radon levels and cancer incidence [29].

Influence of environmental factors including radon on cancer development among population of North and East Kazakhstan regions were investigated by group of researchers from North Kazakhstan State University. They found a positive association between residential radon concentration in these two regions of Kazakhstan and different cancer diseases, predominating lung cancer [29].

7. Health effect of radon

Alpha radiation in chronic exposure can cumulate the biological effect. In comparison with the damaging effects of beta and gamma radiation, alpha particles cause 23–43 times more severe radiation damage. The long-term after effect of the radon is its high blastomogenic activity.

The role of radon and its radioactive decay products as human carcinogens has been established by the International Agency for Research on Cancer in 1988 and is supported by experimental evidences obtained in laboratory animals [30] and epidemiologic studies in humans [31]. In 1990, Schoenberg et al. [32] published a report on the strong association between indoor radon exposure and lung cancer risk. In 2002, Wang et al. [33] has shown that high levels of residential radon increase the risk of lung cancer. Some studies demonstrated that lung cancer risk increases with increasing indoor radon concentration. As reported by Darby et al. [34], cumulative risk of lung cancer when reaching the age of 75 years is estimated to never smokers at 0.4, 0.5, and 0.7% for the radon concentrations of 0, 100, and 400 Bq/m^3, respectively. While cumulative risk of lung cancer at the age of 75 years for smokers reaching 10, 12, and 16% for radon concentrations of 0, 100, and 400 Bq/m^3, respectively. Baysson et al. [35] found that the risk of lung cancer increases from 4 to 28% with increasing radon levels per 100 Bq/m^3. Moreover, the risk of cancer increases in smokers group from 2 to 80% with an increasing radon levels per 100 Bq/m^3.

Specifically, radon emerged as a prominent non-tobacco carcinogen strongly associated with lung cancer. Although the initial evidence supporting the association of radon and lung cancer came from studies involving mine workers, in recent years research has focused on indoor Radon exposure and its risk to the general population. In 2009, the WHO identified chronic residential exposure to radon and its decay products as the second cause of lung cancer after tobacco consumption and as the main risk-factor in never smokers [36]. Torres-Durán et al. [37] studied 192 lung cancer cases and 329 controls in Galicia, Spain. The study subjects were male or female never smokers aged over 30 years. According to their study, adenocarcinoma was the most common histologic type of lung cancer.

In the literature, there are different data on the histological types of lung cancer induced by radon. Some studies have reported that radon increases the risk of small cell lung cancer. Barros-Dios and colleagues [38] found that less frequent histologic types (including large cell carcinomas), followed by small cell lung cancer, had the highest risk associated with radon exposure. According to the study of Taeger et al. [39], a small cell lung cancer and squamous carcinoma is associated with the radon exposure, but adenocarcinoma is associated less. Some studies have found a positive correlation between the incidence of adenocarcinoma in the group of non-smoking women and the increasing concentration of radon in the living room [40]. Wilcox et al. [41] failed to show a statistically showed that radon exposure had a strong effect on small cell lung cancer in cases of both genders, but the causative relation of indoor radon and squamous cell carcinoma was only demonstrated in male cases. Some studies have showed evidence that indoor radon exposure is strongly related to small cell carcinoma and squamous cell carcinoma of the lung.

It is considered worldwide that 3–20% of all lung cancer deaths are caused by indoor radon exposure [42]. The highest estimate of percent risk of lung cancer deaths, which consist 20% from whole mortality of lung cancer, was reported in the study for an average radon concentration of 110 Bq/m^3 [43]. The lowest indoor radon concentration (21 Bq/m^3) was measured in the UK, which also showed the lowest percent risk (3.3%) of lung cancer deaths [44].

The recent studies in Spain have shown a statistical association between indoor radon and lung, stomach, and brain cancer in women [45]. Thus, the effect of radon is not limited to an increased risk of lung cancer, but it can also lead to the development of pathological changes in other organs. The results of mortality studies in North Carolina showed that groundwater radon is associated with increased risk not only for lung but also for stomach cancer [46]. Barbosa-Lorenzo et al. also observed a strong and significant association between residential radon and the incidence of stomach cancer, but their findings were limited by a low number of cases [47]. There are very few studies on other cancers different than lung and stomach cancer.

Bräuner et al. [48] found significant associations between long-term exposure of radon to Danish population and risk of primary brain tumors. Exposure of ionizing radiations has been shown to increase the risk of central nervous system tumors in children and adolescents exposed to computed tomography scans [49]. Some epidemiological studies have suggested a positive correlation between environmental radon exposure and prostate cancer [50]. A study in Galicia (Spain) showed a correlation between brain cancer mortality and indoor radon exposure [51].

Li and colleagues' study demonstrated the evidence of malignant transformation of human benign prostatic epithelial cells exposed to a single dose of alpha radiation [52]. The results of study for breast cancer in Spain shown a certain association between breast cancer risk and radon exposure might be present [45]. However, Turner et al. did not find any association for radon exposure and breast cancer mortality [53].

Recent study observed a significant positive association between mean county-level residential radon concentrations and chronic obstructive pulmonary disease mortality [54].

The risk for bronchopulmonary system following inhalation of radon is significantly higher than for other organs. A detailed study of lung cancer showed that due to the peculiarities of the movement of the mucus and the depth of the location of the basal cells, there are some "risk cells" that receive the highest dose of alpha radiation on exposure to radon.

From the molecular viewpoint, the mechanism of lung cancer induction by radon is not known in detail, but involves both genetic and epigenetic pathways [55, 56].

Because radon is inert, nearly all of the gas that is inhaled is subsequently exhaled. However, ^{222}Rn decays into a series of solid short-lived radioisotopes depositing within the respiratory tract. Because of its relatively short half-lives, the radon-progeny mainly decay in the lung before clearance can take place. Two of these short-lived progeny, Polonium-218 and Polonium-214, emit radioactive alpha particles that, impacting pulmonary epithelium, may genetic-molecular alterations that are mutagenic and result in an increased risk of carcinogenesis [57].

8. Health effect of radon on the population of Kazakhstan

Epidemiological studies show that age-standardized cancer rate in former NIS countries (New Independent States), including Kazakhstan, doubled during the last 20 years. Release of radioactive and toxic debris into the environment could contribute to such an increase.

During the last decades, contamination by industrial or military wastes in Kazakhstan may have contributed to the observed health problems. One of the most heavily polluted areas in Kazakhstan, where environmental pollution exceeds permissible standards, is the East Kazakhstan province. In this area mining and smelting industries and metallurgic complexes are extensive. About half of solid toxic waste (49.2%) accumulated in Kazakhstan are located around enterprises. Surveys of environment exposure to genotoxic agents, including uranium and its short-lived radioactive alpha- and beta-emitting particles, show that environmental health problems in this region are critical, due to the industrial activity of different metallurgic complexes and accumulation of high concentrations of hazardous toxicants.

For the first time in Kazachstan, complex studies on the effect of radon on the incidence disease were conducted in the Zhezkazgan region [58]. In 1985, the high level of radon in the village of Akshatau was detected. The morbidity of 2166 inhabitants of all social and age groups of this settlement was studied. From 1986 to 1990 average morbidity was 19,927 cases per 10,000 of the population, which was 2.9 times higher than in Zhezkazgan region.

The highest percentage of all diseases was accruing to respiratory diseases (62.2%). The second place was occupied by diseases of the nervous system and sense organs (9.3%), the third place—diseases of the circulatory system (7.1%). Analysis of the morbidity structure showed that people living in houses with a high indoor radon level often sought medical advice, especially with symptoms of bronchopulmonary diseases. These cases of seeking medical advice accounted for 63.1% of all diseases and 66.2% of bronchopulmonary diseases. The highest morbidity was observed among children who lived in houses and visited kindergartens functioning in buildings with high indoor radon level (48,625 cases per 10,000 of the population of this age), which 4.6 times higher comparing with children who living in homes with low radon levels. A significant difference in the incidence of respiratory organs was observed in newborns living in houses built using waste from uranium production. The morbidity reached 40,866 cases per 10,000 populations, accounting for 81.1% of the total respiratory morbidity in children of this age. About 65.6% of seeking medical advice among children of school age due to respiratory diseases occurred in those living in houses with a high level of radon and attending school, in which the level of natural gamma background exceeded the permissible level 7.7 times. The most common forms of respiratory diseases in the children, described above, were acute bronchitis, pneumonia and acute infections of the upper respiratory tract. In the adult population of the Akshatau village the highest morbidity level was among people working with the extraction and processing of uranium-bearing ore. Among morbidity of adults were most common chronic and acute bronchitis, chronic diseases of tonsils and adenoids, chronic pharyngitis and nasopharyngitis, acute laryngitis and tracheitis, pneumonia and emphysema [58].

Scientific-research Institute for Radiation Medicine and Ecology of Semey in 2013–2014 conducted a study of radon symptoms in the Kalachi village of Akmola region because of the increase of cases of so-called "Kalachi syndrome" among residents of the village. Of the 59 investigated buildings (residential and social significance) 20% of the buildings' mean values of radon volume activity do not meet the requirements of existing regulations. There is a relationship of "Kalachi syndrome" with points of high radon concentrations. In addition to the risk of developing lung cancer should be noted the property of radon as inert gas to exhibit an anesthetic effect [59].

In accordance with the Decree of the Government of the Republic of Kazakhstan No. 34 of 13-01-2004, "On the approval of the list of diseases associated with exposure to ionizing radiation and the Rules for Determining the Causation of Diseases from Exposure to Ionizing Radiation" the following list of diseases associated with exposure to ionizing radiation was approved:

1. Acute leukemia (leukemia)
2. Malignant lymphomas
3. Solid malignant neoplasm
4. Lung cancer
5. Thyroid cancer
6. Breast cancer
7. Urologic cancer
8. Gastric cancer (including oral, pharynx, esophagus, stomach and colon cancer).

The analysis showed that the population of Kazakhstan lives primarily in potentially dangerous areas from the radioactivity exposure point of view. According to studies, level of radon in premises where people stay for long times may exceed allowable levels by several times. Accordingly, the study of the risks of radon-induced diseases, and primarily lung cancer, is necessary.

9. Lung cancer in Kazakhstan

The incidence rates of malignant diseases in various areas vary quite widely (**Table 1**). It can be noted that the average number of diseases per 100,000 population in regions with an increased level of radon hazard (North Kazakhstan, Pavlodar, East Kazakhstan, Kostanay, Karaganda, Akmola, and Almaty) is 276, while the average number of diseases per 100,000 population by regions with a normal level of radon hazard (Aktobe, Atyrau, Zhambyl, West Kazakhstan, Kyzylorda, Mangystau, and South Kazakhstan)—145.5.

The data on cancer incidence by regions are given in **Table 1** and in **Figure 2**. A comparative analysis of this map and a map of radon-affected areas (**Figure 1**) indicates a correlation between the incidence of cancer and the level of radon hazard.

Of course, for this analysis, more detailed studies are needed. For example, the areas of South Kazakhstan, Almaty, and parts of Kyzylorda regions are classified as potentially radon-affected

http://dx.doi.org/10.5772/intechopen.71135

areas because they are located in the territories of uranium provinces. However, these areas are characterized by low cancer rate. This is explained by the fact that unlike typical uranium provinces (e.g. the North Kazakhstan province), uranium deposits in these areas are located at great depths and do not surface.

However, when comparing the levels of radon and the cancer rate, there is a sufficient correlation, indicating an undoubted dependence of oncological morbidity and radon hazard. This dependence is complex and requires a more detailed study, on the one hand, of various factors affecting the health of the population: chemical pollution, smoking, etc., and on the other hand— natural geological features of the territory.

It is possible to distinguish two groups of sources of radon intake from the subsoil. Firstly, the sources are the rocks themselves with a high geochemical background, which can have a radon concentration in the ground up to 50–100 Bq/l, which can form radon-affected areas. Secondly, the source is radon-bearing tectonic zones with well-defined linear parameters. The level of radon in the air of houses located in such zones can reach very high values, reaching 10 thousands of Bq/m^3.

The frequency of lung cancer by region almost completely coincides with the incidence of cancer.

Region	2012	2013	Rate of increase (%)
Republic Kazakhstan	190.6	193.9	1.7
Akmola	233.6	248.5	6.4
Aktobe	173.9	179.1	3.0
Atyrau	124.4	140.5	13.0
East Kazakhstan	276.2	285.5	3.4
Zhambyl	129.5	139.9	8.1
West Kazakhstan	207.5	207.3	–
Karaganda	257.1	258	0.4
Kyzylorda	139.6	138	–
Kostanay	291.9	278	–
Mangystau	113.9	123.6	8.6
Pavlodar	301.5	301.9	0.1
North Kazakhstan	309.2	312.5	1.1
South Kazakhstan	93.3	98.6	5.7
Astana	169.2	172.3	1.8
Almaty	245.2	247.9	1.1

Table 1. Cancer morbidity in Kazakhstan (per 100,000 people).

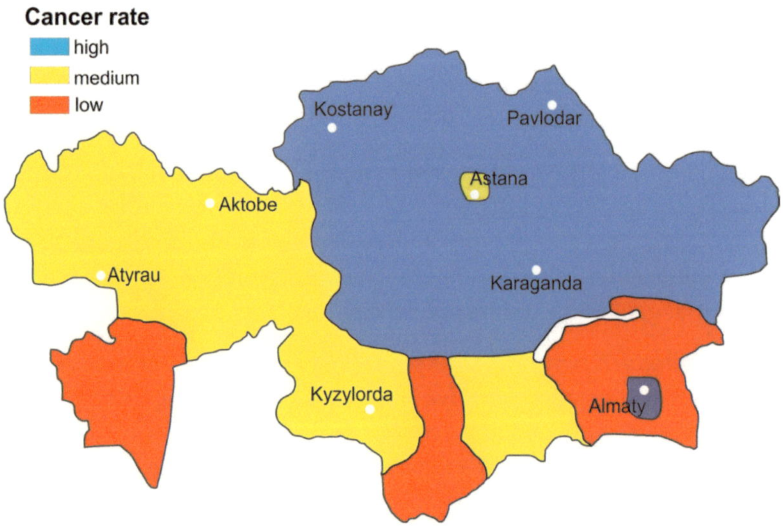

Figure 2. Cancer rate in the Republic of Kazakhstan.

In the structure of the incidence of cancer lung ranks first place in Akmola, Aktyubinsk, Atyrau, Kostanay, Pavlodar, and North Kazakhstan regions, the second place in East Kazakhstan, Zhambyl, West Kazakhstan, Mangystau regions, and Astana city, the third place in Karaganda, South Kazakhstan, and Almaty.

Lung cancer in the structure of oncology occupies the second rank position, and its share at the moment was 11.4%. This is the most widely distributed form of cancer which leading since 1985 in Republic. The highest morbidity of lung cancer is observed in the North Kazakhstan, Pavlodar, and Akmola regions.

When analyzing the morbidity of lung cancer in the East Kazakhstan region, the areas with a higher lung cancer rate (more than 40/100,000 of population) are clearly associated with a radon-affected territory in the northern part of the region. In 15 settlements of 18 examined in this area, the excess of the norm (sometimes significant—up to 40 times) of radon concentration was established.

The area with increased oncological morbidity in the northeast of the East Kazakhstan region borders on the territory with an increased incidence in the Altai territory (Russia) [58].

However, the genotoxic effects of radon on population of Kazakhstan are poorly understood, in spite of the fact that many regions of the country contain the high levels of radon. Studies elucidating potential health risk among population exposed to radon and genotoxic effect of radon in Kazakhstan are very limited or these studies have never conducted in some areas.

10. Genetics and epigenetic effects of radon

Alpha particles interact with DNA either directly or indirectly through the generation of free radicals, leading to double-strand breaks, large chromosomal aberrations, and point mutations. In a previous cytogenetic study conducted on a population residing near uranium tailings and mining activities in North Kazakhstan, we have demonstrated an increased frequency of chromosomal aberrations in uranium miners compared to matched control subjects, who were not exposed to uranium compounds or to any other known chemical or physical mutagens. The data on aberration type frequency showed a predominance of aberrations of chromosome type in the exposed group, mainly including paired acentric fragments, dicentric chromosomes, and centric rings [60].

Several investigations on chromosome aberrations of blood lymphocytes due to radon exposure were previously carried out. In a study by Hellman et al. with human blood samples to detect DNA damage, an increased concentration of radon in indoor air was found to be associated with enhanced DNA damage in peripheral lymphocytes [61]. The findings of Li's study demonstrated that DNA damage in peripheral blood mononuclear cells was significantly increased in a dose-dependent manner [62]. Abo-Elmagd et al. [63] showed the formation of different types of structural chromosomal aberrations, where chromosomal and chromatid breaks had the dominant incidence. Moreover, damaging effect of radon in the blood and bone marrow was expressed as a reduction in the mitotic index and an increase in chromosome damage. The average frequencies of single and double fragments, chromosomal exchanges, total number of aberrations, chromatid type, chromosome type, and all types of aberrations were significantly increased in the radon-exposed group [64]. As reported by several authors, there is a synergistic effect of smoking and radon on increasing the frequency of chromosomal aberrations. Meenakshi and Mohankumar showed that the frequency of dicentrics in radon-exposed smoker cells was found to be higher than non-smokers by factor of 3.8 [65]. Moreover the comparison of three groups: miners, workers of uranium enterprises, and a control group revealed that chromosomal aberrations were a consequence of the effect of radon, but not of uranium [66]. Disruptions to normal chromosomal arrangement represent a major contribution to carcinogenesis. Cytogenetic analysis, including the detection of chromosomal aberrations, has been used for many decades as a tool to determine mutagenic and carcinogenic risks. Chromosomal aberrations (insertions, deletions, translocations, ring formation, duplication, and inversions) as a result of radon exposure significantly increase the risk of developing neoplastic processes and enhance carcinogenic ability and this can be one of the mechanisms of development of radon-induced lung cancer.

In addition to chromosomal aberrations, gene mutations play an important role in the pathogenesis of lung cancer. Specific "hotspot" mutations in cancer-relevant genes have been described in radon-induced lung cancer. Sequencing of the TP53 tumor-suppressor gene has provided links between specific mutations and radon exposure. TP53, also known as the "guardian of the genome" owing to its involvement in regulation of the cell cycle and apoptosis upon DNA damage, plays a critical role in the cellular response to genetic damage caused by radiation. Most papers on TP53 mutations in radon-associated lung cancer consist of occupational studies in uranium miners [67, 68]. These studies showed a TP53 mutational spectrum different from those seen in lung cancers caused by tobacco smoke. Vähäkangas et al. showed that most frequent base substitutions associated with tobacco smoking (G:C to T:A transversions) were not identified in uranium miners [67]. The study of Popp et al. indicated an overrepresentation of codon 213/3 polymorphism in p53 in radiation-caused cancers [69]. Some studies reported a radon-related TP53 hotspot in codon 249, exon 7. These mutations, mostly transversions and small deletions very uncommon in smoke-induced lung cancers, were mainly identified in squamous cell carcinomas (SCC) and in large cell carcinomas. In the same cohort of miners, the presence of codon 249 transversion was not demonstrated in lung adenocarcinomas, thus implicating tumor histology as determinant for this specific mutation [70]. Only few studies analyzed TP53 punctual mutations in lung tumors from residential radon-exposed individuals and their results are not univocally supportive of the mentioned hotspot in codon 249 [71, 72]. However, Hollstein et al. proposed that loss of tumor-suppressor function of TP53 due to α-particle radiation may occur preferentially by mechanisms such as intrachromosomal deletions, rather than by base substitution mutations [73].

Accordingly, single nucleotide polymorphisms (SNPs) in TP53 sequence may affect lung cancer susceptibility by influencing hundreds of target genes. Nevertheless, the mechanistic details underlying the molecular pathways by which TP53 gene polymorphisms are associated with radon-induced lung cancer remain unclear.

Further investigations are required in order to understand if it is possible to identify hotspot regions for radon-induced lung cancers, providing a unique biomarker that contributes to understand the etiology of the disease and to elucidate the risk as occurring at low exposure levels.

There is growing evidence that the tumor suppressive activity of TP53 is related with miRNA expression. miRNAs are small, non-coding RNAs regulating gene expression at the post-transcriptional level. microRNAs bind to complementary sequences on target messenger RNA transcripts, usually resulting in translational repression or target mRNA degradation and gene silencing.

miRNA profiling after TP53 induction, pointed miR-34a/b/c as the most up-regulated miRNAs identifying miR-34 as a main TP53 gene effector [74]. The expression of miR-34 family members is induced after genotoxic stress in a p53-dependent manner in cultured cells as well as in irradiated mice [74–76]. Notably, overexpression of miR-34a increases apoptosis, whereas miR-34a inactivation strongly attenuates p53-mediated apoptosis in cells exposed to genotoxic stress. Further miRNAs, including miR-125a, have recently been linked to p53-regulated apoptosis in lung cancer cells [77]. Several TP53 mutants, linked to oncogenic progression, suppress post-transcriptional maturation of miRNAs bearing growth-suppressive functions [78].

As shown in a number of studies, the profile of microRNA can vary both under the action of chemical agents [79, 80] and radiation [81–86].

Recently, it has been shown that ionizing radiation can induce changes in miRNA expression profiles in normal human fibroblasts [87] and immortalized cell lines [88, 89]. In vivo studies found that miRNA signatures induced by ionizing radiations in mouse blood are dose and radiation type-specific [90] indicating that miRNAs can be exploited as biomarkers of exposure to radiation [91].

Studies in humans also highlighted the importance of miRNAs to evaluate the biological outcomes of radiation exposure. Evidence from a study conducted in patients undergoing radiotherapy, indicated that miRNA expression can be used as biomarker of radiation exposure in humans [92]. Another study examined the utility of miRNA signatures in lung cancer screening using blood samples from computerized tomography trials. The authors performed extensive miRNA profiling of primary lung tumors, paired normal lung tissues, and multiple plasma samples collected before and at the time of the disease. This approach was able to define a plasma miRNA signature associated with increased risk of lung cancer [93]. Also other groups have integrated miRNA profiles into computerized tomography-screening programs [94, 95].

A number of studies have shown the radioprotective role of some types of microRNA. In vitro studies using the human diploid fibroblast cell line WI-38 demonstrated that the mature form of the miR-155 inhibits premature "cell aging" induced by radiation [96].

In this regard some scientists make a supposition that some microRNAs can determine the resistance of tumor cells to radiotherapy and be a prognostic biomarker for monitoring of cancer treatment strategy.

Ma et al. demonstrated that overexpression of miR-622 in colon cancer cells inhibits Rb by directly binding to RB1-3'UTR, thus inactivating the Rb-E2F1-P/CAF complex, which is responsible for the activation of pro-apoptotic genes in response to the effect of radiation [97].

On the other hand, many studies have shown that overexpression is one of the most common types of cancer. A study of 48 patients diagnosed with cervical cancer revealed the effect of miR-18a function in the regulation of cellular radiosensitivity. Liu et al. showed that miR-18a suppresses the activity of ATM, reduces the repair of double-strand breaks and stimulates apoptosis of irradiated cancer cells [98].

Similar results were obtained by Lan and colleagues in their study of non-small cell lung cancer. According to their results, miR-15 and miR-16 increase the radiosensitivity of H358 and A549 cell lines and the level of apoptosis in vitro [99].

Thus, the level of expression of different types of microRNA can react differently to the effect of ionizing radiation [100]. Most likely the effect of radiation depends on target genes inhibited by a specific microRNA. However, according to numerous literature data, it is impossible to identify a clear separation of microRNAs with respect to radioactive effects. Expression of most of the currently known microRNAs can be amplified or suppressed by ionizing radiation depending on the type of cells of their producing [100].

Other researchers believe that a change in the level of microRNA expression under the influence of radiation has a dose-dependent effect [101]. Lee et al. showed that the greatest effect on the profile of microRNAs in the mononuclear phagocytes of human blood exposed to different doses of radiation had 1 Gy irradiation, while 0.5 and 2.5 Gy irradiation showed very slight changes in the level of microRNA expression [101]. In order to explain the results obtained, the authors compared the biological effect in the cells and target genes of miRNAs sensitive to radiation. It turned out that the majority of these microRNAs were involved in the control of homeostasis, neuronal survival, and regulation of the cell cycle and proliferation. Thus, sensitive to radiation microRNA (including the action of small radiation doses) can be involved in the pathogenesis of cancer caused by the action of radiation. Moreover, Japanese scientists from the University of Hiroshima, after conducting a study of patients diagnosed with stomach cancer exposed to large and small doses of radiation, found that the miR-24 expression profile was largely changed in the first group. Based on the results obtained, Naito et al. suggest that miR-24 can be use as a biomarker risk of developing gastric cancer induced by radiation [102].

Despite the large number of works devoted to the study of the relationship between microRNA, radiation, and carcinogenesis, there are few studies about the role of microRNA in radon-induced lung cancer. Chen et al. reported that the level of expression of let-7 microRNA in human bronchial epithelial (HBE) cells due to radon exposure is decreasing, which results in activation of the K-ras oncogene. Furthermore, a significant down-regulation of K-ras was then confirmed in both let-7b-3p and let-7a-2-3p transfected HBE cells [103]. The results of this study proposed that let-7 microRNA and K-ras are involved in radon-induced lung damage both in vivo and in vitro. In vitro study of malignant transformation due to radon exposure showed the change in the expression level of a number of microRNAs, including hsa-miR-483-3p, hsa-miR-494, hsa-miR-2115*, hsa-miR-33b, hsa-miR-1246, hsa-miR-3202, hsa-miR-18a, hsa-miR-125b, hsa-miR-17*, and hsa-miR-886-3p. These miRNAs target genes play key role in the regulation of cell proliferation, differentiation, adhesion during the process of malignant transformation, which are associated with different signal pathways such as mitogen-activated protein kinase (MAPK), Int and Wg (Wnt), reactive oxygen species (ROS), and nuclear factor κB (NF-κB) [104]. Recent study showed an increase in the expression of some circular RNA (class of endogenous non-protein coding RNAs that contain a circular loop) in mouse lung tissues due to radon exposure and 5 microRNAs binding sites detected for each circRNA. The authors suggest that these circRNAs and miRNAs could play important and synergistic roles in radon-induced lung damage, even in lung cancer [105].

Thus, the role of microRNA in the pathogenesis of radon-induced lung cancer is still unclear. Nevertheless, the search for biomarkers for early diagnosis of lung cancer, which can be useful, especially for Kazakhstan, since most of the Republic's territory has a high natural level of radon.

At the epigenetic level, not only miRNAs expression patterns, but specific alterations including alterations in DNA methylation, histone modifications leading to generation of γ-H2AX species (indicative of double-strand DNA breaks) have been associated with radon exposure [106].

Masoomi et al. showed that people living in regions with a high natural background of radiation had faster DNA repair [107]. Earlier, Fernandez-Capetillo et al. was proposed that H2AX phosphorylation may play an important role in the low-dose radiation adaptive response [108]. H2AX phosphorylation-induced chromatin restructuring for DNA repair/signaling factors and/or tether DNA ends. If this process is induced by low-dose radiation and if there is a memory for mounting this process in the event of a later enhanced genomic stress (e.g. high dose of carcinogen), more rapid and more efficient DNA repair might be expected to occur [109].

DNA methylation is well-described in lung cancer and is considered to be potentially reversible [110]. There is much less information about the effect of radon on the status of DNA methylation. In study of Chinese uranium workers has shown the methylation of O-6-methylguanine-DNA methyltransferase (MGMT) and p16 gene in the sputum cells is related to the accumulate exposure dosage of the radon [111]. p16 is a tumor suppression gene, which plays key role in cell cycle regulation and methylation p16 gene is significantly associated with non-small cell lung cancer (NSCLC) [112]. O-6-methylguanine-DNA methyltransferase is a DNA repair enzyme which protected cells from the carcinogenic effect of alkylating agents. The hypermethylation of the MGMT gene's promoter may play a significant role in lung carcinogenesis [113]. Belinsky et al. has shown that exposure to plutonium, which similar to radon exerts its effects through alpha particles, can induce p16 gene inactivation by promoter methylation [114].

It is known, that radon is the first reason of lung cancer in never smoking population. But some studies showed that higher methylation rate in p16 is dependent on the amount of exposure to tobacco smoke and this epigenetic change is specific to smokers with lung cancer [115]. By comparison in non-smokers patients with lung cancer, the methylation rate in p16 is low [116]. Bastide et al. concluded that, in radon-induced rat lung tumors, promoter methylation did not affect the p16 gene inactivation although the lack of p16 protein expression in lung tumors remains to be established [117].

The study in United States showed exposure to radon through uranium mining did not affect the prevalence for MGMT methylation [118].

Perhaps, in this case we are talking about the dose-dependent effect of radon on epigenetic reprogramming. High-dose radiation can promote epigenetically silencing of adaptive-response genes, while low doses of natural radiation can, on the contrary, have a protective effect and activate the same genes through changes at the epigenetic level. More importantly, according to the Environmental Protection Agency low-level radon exposure (4 pCi L − 1) may actually be protective against lung cancer [119]. Some study demonstrated cancer prevention after low-level radiation exposure [120, 121]. For this reason it will require more research to reconcile these contradictory data.

There is not much information about the effect of radon exposure on histone modification. The study by Pogribny has shown the effect of low-dose radiation exposure on histone H4-Lys 20 trimethylation in the thymus tissue using an in vivo murine model. In such a case, the loss of histone H4-Lys20 trimethylation was accompanied by a significant increase in global DNA hypomethylation and reduced expression of DNMT1 and DNMT3b. Moreover, the levels of methyl-binding proteins MeCP2 and MBD2 were significantly reduced in thymus of exposed females and males mice compared with respective controls. In such a case, the loss of histone H4-Lys20 trimethylation was accompanied by a significant increase in global DNA hypomethylation and reduced expression of DNMT1 and DNMT3b. Moreover, the levels of methyl-binding proteins MeCP2 and MBD2 were significant decreased in thymus of exposed females and males mice compared with respective controls [122].

The studies of specific molecular mechanisms that can lead to genetic and epigenetic aberrations in lung tumor genomes as a result of exposure to radon can give a unique opportunity for further developing diagnostic and identification of candidates for therapeutic targets. However future studies are needed to explore both the genetic and epigenetic mechanisms in pathogenesis of cancer and fully elucidate the effects of radon on health and disease.

11. Lung cancer studies in Kazakhstan

According to the latest WHO data published in May 2014, lung cancers deaths in Kazakhstan reached 3836 or 2.58% of total deaths. The age adjusted death rate is 26.24 per 100,000 of population ranks Kazakhstan #36 in the world [123]. Currently, Kazakhstan is in need of advanced actions to reduce lung cancer rate. However, unfortunately, in Kazakhstan there are very few studies of the genetic and epigenetic mechanisms of the lung cancer pathogenesis, especially the radon-induced action.

Sagyndykova studied the cellular and molecular carcinogenic effects on lung tissue due to radiation exposure on the population, lived on the territories of Semipalatinsk region. These territories were used for testing nuclear weapons for the Soviet army during the period 1949–1989. The results of the study showed a higher level of expression of IGFBP1, IGFBP2, p53, c-mus, bax and reduced expression of IGFBP3, IGFBP4, IGFBP5, bcl-2 in tumor cells compared to the control group (patients with lung cancer not due to radiation). Probably there are some differences in the mechanisms of proliferation, differentiation, and apoptosis in radiation-induced lung tumors [124]. We have long been studying the effect of radiation in the population of Kazakhstan [125]. In our previously study, we showed the association between the Gln/Gln genotype of gene XRCC1 and the frequency of chromosomal aberrations in the uranium workers. The frequency of chromosomal aberrations in heterozygous carriers of the XRCC3gene Thr/Met was lower than in the homozygous carriers of the wild type Thr/Thr [126]. It is known that DNA repair genes are involved in the formation of individual sensitivity to radiation exposure. Larionov et al. showed that three SNP of DNA repair genes may be considered as radon-sensitive markers [127]. Considering the fact, that the association between the DNA repair genes polymorphisms and radon exposure was found, similar studies indicate a prospect of searching for markers of individual sensitivity to radon among allelic variants of key DNA repair genes.

As discussed previously in the review, in addition to genetic factors, the epigenetic changes play an important role in the pathogenesis of lung cancer, including radon-induced lung cancer. One of the goals of contemporary cancer research is the development of new markers that facilitate earlier and non-invasive diagnosis. Studies have shown that miRNAs expression levels are altered in cancer. Recently, extra-cellular miRNAs have been detected in biological fluids and studied as possible cancer markers that can be detected by non-invasive procedures. The serum concentrations of several microRNAs have been associated with lung cancer [128]. Moreover, miRNAs have been reported to play an important role in inducing resistance to anti-cancer drugs [129]. miRNAs can be exploited not only as biomarkers of radiation exposure but also as biomarkers of lung cancer. The problem of the health risk exposed to radon is not sufficiently studied on the populations of Kazakhstan, although the country (mainly its East and North areas) has the regions containing the high levels of radon [10]. Currently, we aim at addressing the role of genetic and epigenetic factors affecting susceptibility to radon exposure. At epigenetic level, we investigate the effects of radon on the expression of miRNAs in the blood plasma of exposed subjects. This will allow us to validate circulating miRNAs as a reliable biomarker of early biological effect to be used prior to the development of lung cancer in a high-risk population. Particularly, it is very promising to develop highly specific and efficient test systems of lung cancer based on molecular markers, which levels the change due to radon exposure.

Author details

Rakhmetkazhy Bersimbaev* and Olga Bulgakova

*Address all correspondence to: ribers@mail.ru

L.N. Gumilyov Eurasian National University, Institute of Cell Biology and Biotechnology, Astana, Kazakhstan

References

[1] Berikbolov BP, Petrov NN, Karelin VG. Uranium Mine Ore Resources of Kazakhstan. Almaty: Nauka Press; 1996. 220 p (in Russian)

[2] Stegnar P, Shishkov I, Burkitbayev M, Tolongutov B, Yunusov M, Radyuk R, Salbu B. Assessment of the radiological impact of gamma and radon dose rates at former U mining sites in Central Asia. Journal of Environmental Radioactivity. 2013;**123**:3-13. DOI: 10.1016/j.jenvrad.2012.12.005

[3] Salbu B. Preface: Uranium mining legacy issue in Central Asia. Journal of Environmental Radioactivity. 2013;**123**:1-2. DOI: 10.1016/j.jenvrad.2011.12.010

[4] Meenakshi C, Sivasubramanian K, Venkatraman B. Nucleoplasmic bridges as a biomarker of DNA damage exposed to radon. Mutation Research. 2017;**814**:22-28. DOI: 10.1016/j.mrgentox.2016.12.004

[5] Kazymbet PK. Radioecological state of the residential areas in the uranium mining regions of Kazakhstan. Scientific Proceedings of Institute for Radiobiology and Radiation Protection. 2014;**1**:19-55 (in Russian)

[6] Environment and Sustainable Development in Kazakhstan. A Series of UNDP Publication in Kazakhstan. №UNDPKAZ06. Almaty; 2004 (in Russian)

[7] Kazatomprom. National Atomic Company [Internet]. 2015. Available from: www.kazatomprom.kz [Accessed 15-09-2017]

[8] Zoriy P, Ostapczuk P, Dederichs H, Höbig J, Lennartz R, Zoriy M. Biomonitoring of environmental pollution by thorium and uranium in selected regions of the Republic of Kazakhstan. Journal of Environmental Radioactivity. 2010;**101**:414-420. DOI: 10.1016/j.jenvrad.2010.02.014

[9] The Decree of the Government of the Republic of Kazakhstan on the approval of the program №1006. Supplemental legal system. Almaty; 2002 (in Russian)

[10] Bersimbaev RI, Bulgakova O. The health effects of radon and uranium on the population of Kazakhstan. Genes and Environment. 2015;**37**:18. DOI: 10.1186/s41021-015-0019-3

[11] Salbu B, Stegnar P, Strømman G, Skipperud L, Rosseland B, Heier L, Lind O, Oughton D, Lespukh E, Uralbekov B, Kayukov P. Legacy of Uranium Mining Activities in central Asia - Contamination, Impact and Risks.UMB Report. Draft Project Report of Results Obtained within the NATO RESCA Project and the Joint Project between Norway, Kazakhstan, Kyrgyzstan and Tajikistan. Norwegian University of Life Sciences; 2011. 156 p. DOI: 10.13140/RG.2.1.3814.8562

[12] Fedorov M, Kayukov P, Bensman V. The study of the relationship between the concentration of soil radon and its content in the indoor air, the development of the criteria for assessing the areas according to the degree of radon hazard. In: Ministry of Environmental Protection of the Republic of Kazakhstan. Limited liability partnership "Ecoservice-C". Final report. Astana; 2011 (in Russian)

[13] Fedorov M. Conducting radiation monitoring in the rural settlements. In: Report 2008-2011 in 4 books. Almaty; 2011 (in Russian)

[14] Soroka Y, Molchanov A. Radiation and radon survey of Akchatau (Khazakstan) and experience with radon remedial measures. Radiation Protection Dosimetry. 1998;**78**(3):231-236. DOI: 10.1093/oxfordjournals.rpd.a032356

[15] Sevostyanov VN. Some aspects of the radon problem in Kazakhstan. Radioactivity in the Environment. 2005;**7**:409-419. DOI: 10.1016/S1569-4860(04)07047-0

[16] Kobal I, Vaupotič J, Gregorič A, Uralbekov B. Comparison of approaches in Slovenia and Kazakhstan in managing exposure to radon. In: Merkel BJ, Arab A, editors. Uranium—Past and Future Challenges. Switzerland: Springer International Publishing; 2015. p. 689-698. DOI: 10.1007/978-3-319-11059-2_79

[17] Efremov GF. Hydrogeological Survey of Aquifers to Assess Radionuclide Contamination. Almaty: Volkovgeologiya; 2001. 200 p (in Russian)

[18] The Ministry of healthcare and social development of the Republic of Kazakhstan. [Internet]. 2014. Available from: https://www.mzsr.gov.kz [Accessed: 15-09-2017]

[19] Klodzinskiy A, Bekenova F, Baydurin S. Features of respiratory function in uranium operators workers suffering from chronic obstructive pulmonary disease. Astana Medical Journal. 2005;**3**:111-113 (in Russian)

[20] Djumasheva R, Kazymbet P. Effect of prolonged exposure of uranium industrial dust to the lungs of humans and animals. Scientific Proceedings of Institute for Radiobiology and Radiation Protection. 2014;**1**:146-199 (in Russian)

[21] Bekenova F, Baydurin S, Klodzinskiy A, Zhautikova B. The prevalence of diseases of the internal organs in the cohort of the workers Stepnogorsk Mining Factory. Clinical Medicine of Kazakhstan. 2006;**2**:16-18 (in Russian)

[22] Ahmedyanova H, Ivashevskaya R, Ibrayeva Z. Influence of anthropogenic factors on the course of pregnancy and childbirth in women living in the uranium mining regions. Astana Medical Journal. 2005;**3**:89-90 (in Russian)

[23] Raisova KA. Analysis of the children incidence in uranium mining regions of Akmola. Astana Medical Journal. 2005;**3**:129-131 (in Russian)

[24] Kakabaev AA, Sharipov IK, Bersimbaev RI. A cytogenetic study of Kazakhstan uranium mine workers. Russian Journal of Cytology. 1998; **41**(3,4):274 (in Russian)

[25] Kakabaev AA, Sharipov IK, Bersimbaev RI. Chromosome aberrations in the group of uranium mine workers in North Kazakhstan. News of the National Academy of Sciences of Kazakhstan. (Ser. Biology). 1999;**2**(212):15-21 (in Russian)

[26] Toksobayeva GA, Kakabayev AA, Bersimbaev RI. Investigation of chromosomal aberration frequencies and glutathione-s-transferase M1 and T1 genes in workers occupationally exposed to uranium in northern Kazakhstan. NATO Security through Science Series E, Human and Societal Dynamics. 2010;**73**:177-182. DOI: 10.3233/978-1-60750-645-4-177

[27] Kreuzer M, Grosche B, Schnelzer M, Tschense A, Dufey F, Walsh L. Radon and risk of death from cancer and cardiovascular diseases in the German uranium miners cohort study: Follow-up 1946-2003. Radiation and Environmental Biophysics. 2010;**49**(2):177-185. DOI: 10.1007/s00411-009-0249-5

[28] Nurgaziyev SHK, Seytkazina GD. The frequency of lung cancer in Kazakhstan for 10 years. Siberian Journal of Oncology. 2013;**2**:54-55 (in Russian)

[29] Beletskaya N. Effect of Environmental Factors on Cancer Incidence in the Population of North Kazakhstan and East Kazakhstan. Petropavlovsk: Kozybayev M.SKGU; 2013. 225 p (in Russian)

[30] Collier CG, Strong JC, Humphreys JA, Timpson N, Baker ST, Eldred T, Cobb L, Papworth D, Haylock R. Carcinogenicity of radon/radon decay product inhalation in rats – Effect of dose, dose rate and unattached fraction. International Journal of Radiation Biology. 2005;**81**:631-647. DOI: 10.1080/09553000500368404

[31] Mc Laughlin J. An historical overview of radon and its progeny: Applications and health effects. Radiation Protection Dosimetry. 2012;**152**:2-8. DOI: 10.1093/rpd/ncs189

[32] Schoenberg JB, Klotz JB, Wilcox HB, Nicholls GP, Gil-del-Real MT, Stemhagen A, Mason TJ. Case–control study of residential radon and lung cancer among New Jersey women. Cancer Research. 1990;**50**(20):6520-6524

[33] Wang Z, Lubin JH, Wang L, Zhang S, Boice JD Jr, Cui H, Zhang S, Conrath S, Xia Y, Shang B, Brenner A, Lei S, Metayer C, Cao J, Chen KW, Lei S, Kleinerman RA. Residential radon and lung cancer risk in a high-exposure area of Gansu Province, China. American Journal of Epidemiology. 2002;**155**(6):554-564

[34] Darby S, Hill D, Deo H, Auvinen A, Barros-Dios JM, Baysson H, Bochicchio F, Falk R, Farchi S, Figueiras A, Hakama M, Heid I, Hunter N, Kreienbrock L, Kreuzer M, Lagarde F, Mäkeläinen I, Muirhead C, Oberaigner W, Pershagen G, Ruosteenoja E, Rosario AS, Tirmarche M, Tomásek L, Whitley E, Wichmann HE, Doll R. Residential radon and lung cancer detailed results of a collaborative analysis of individual data on 7148 persons with lung cancer and 14,208 persons without lung cancer from 13 epidemiologic studies in Europe. Scandinavian Journal of Work, Environment & Health. 2006;**32**(Suppl 1):1-83

[35] Baysson H, Tirmarche M, Tymen G, Gouva S, Caillaud D, Artus JC, Vergnenegre A, Ducloy F, Laurier D. Indoor radon and lung cancer in France. Epidemiology. 2004;**15**(6): 709-716

[36] World Health Organization, Handbook on Indoor Radon: A Public Health Perspective, WHO [Internet]. 2009. Available from: https://www.ncbi.nlm.nih.gov/books/NBK143216/ [Accessed: 15-09-2017]

[37] Torres-Durán M, Ruano-Ravina A, Parente-Lamelas I, Leiro-Fernández V, Abal-Arca J, Montero-Martínez C, Pena-Álvarez C, González-Barcala FJ, Castro-Añón O, Golpe-Gómez A, Martínez C, Mejuto-Martí MJ, Fernández-Villar A, Barros-Dios JM. Lung cancer in never-smokers: A case–control study in a radon-prone area (Galicia, Spain). The European Respiratory Journal. 2014;**44**(4):994-1001. DOI: 10.1183/09031936.00017114

[38] Barros-Dios JM, Ruano-Ravina A, Perez-Rios M, Castro-Bernardez M, Abal-Arca J, Tojo-Castro M. Residential radon exposure, histologic types, and lung cancer risk. A case–control study in Galicia, Spain. Cancer Epidemiology, Biomarkers & Prevention. 2012;**21**(6):951-958. DOI: 10.1158/1055-9965.EPI-12-0146-T

[39] Taeger D, Fritsch A, Wiethege T, Johnen G, Eisenmenger A, Wesch H, Ko Y, Stier S, Michael Muller K, Bruning T, Pesch B. Role of exposure to radon and silicosis on the cell type of lung carcinoma in German uranium miners. Cancer. 2006;**106**(4):881-889. DOI: 10.1002/cncr.21677

[40] Alavanja MC, Brownson RC, Lubin JH, Berger E, Chang J, Boice JD Jr. Residential radon exposure and lung cancer among nonsmoking women. Journal of the National Cancer Institute. 1994;**86**(24):1829-1837

[41] Wilcox HB, Al-Zoughool M, Garner MJ, Jiang H, Klotz JB, Krewski D, Nicholson WJ, Schoenberg JB, Villeneuve PJ, Zielinski JM. Case–control study of radon and lung cancer in New Jersey. Radiation Protection Dosimetry. 2008;**128**(2):169-179. DOI: 10.1093/rpd/ncm330

[42] Kim SH, Hwang WJ, Cho JS, Kang DR. Attributable risk of lung cancer deaths due to indoor radon exposure. Annals of Occupational and Environmental Medicine. 2016;**28**:8. DOI: 10.1186/s40557-016-0093-4

[43] Leenhouts HP, Brugmans MJ. Calculation of the 1995 lung cancer incidence in The Netherlands and Sweden caused by smoking and radon: Risk implications for radon. Radiation and Environmental Biophysics. 2001;**40**:11-21

[44] Radon and public health. Oxford: Health Protection Agency [Internet]. 2009. Available from: http://webarchive.nationalarchives.gov.uk/20101108230210/http://www.hpa.org.uk/web/HPAwebFile/HPAweb_C/1243838496865 [Accessed: 15-09-2017]

[45] López-Abente G, Núñez O, Fernández-Navarro P, Barros-Dios JM, Martín-Méndez I, Bel-Lan A, Locutura J, Quindós L, Sainz C, Ruano-Ravina A. Residential radon and cancer mortality in Galicia, Spain. Science of the Total Environment. 2017;**610-611**:1125-1132. DOI: 10.1016/j.scitotenv.2017.08.144

[46] Messier KP, Serre ML. Lung and stomach cancer associations with groundwater radon in North Carolina, USA. International Journal of Epidemiology. 2017;**46**:676-685. DOI: 10.1093/ije/dyw128

[47] Barbosa-Lorenzo R, Barros-Dios JM, Raı'ces Aldrey M, Cerdeira Carame's S, Ruano-Ravina A. Residential radon and cancers other than lung cancer: A cohort study in Galicia, a Spanish radon-prone area. European Journal of Epidemiology. 2016;**31**:437-441. DOI: 10.1007/s10654-016-0134-x

[48] Bräuner EV, Andersen ZJ, Andersen CE, Pedersen C, Gravesen P, Ulbak K, Hertel O, Loft S, Raaschou-Nielsen O. Residential radon and brain tumour incidence in a Danish cohort. PLoS One. 2013;**8**(9):e74435. DOI: 10.1371/journal.pone.0074435. eCollection 2013.

[49] Mathews JD, Forsythe AV, Brady Z, Butler MW, Goergen SK, Byrnes GB, Giles GG, Wallace AB, Anderson PR, Guiver TA, McGale P, Cain TM, Dowty JG, Bickerstaffe AC, Darby SC. Cancer risk in 680,000 people exposed to computed tomography scans in childhood or adolescence: Data linkage study of 11 million Australians. BMJ. 2013;**346**:f2360. DOI: 10.1136/bmj.f2360

[50] Eatough JP, Henshaw DL. Radon and prostate cancer. Lancet. 1990;**335**(8700):1292

[51] Ruano-Ravina A, Aragones N, Kelsey KT, Pérez-Ríos M, López-Abente G, Barros-Dios JM. Residential radon exposure and brain cancer: An ecological study in a radon prone area (Galicia, Spain). Scientific Reports. 2017;**7**(1):3595. DOI: 10.1038/s41598-017-03938-9

[52] Li H, Gu Y, Miki J, Hukku B, McLeod DG, Hei TK, Rhim JS. Malignant transformation of human benign prostate epithelial cells by high linear energy transfer alpha-particles. International Journal of Oncology. 2007;**31**(3):537-544

[53] Turner MC, Krewski D, Chen Y, Pope CA, Gapstur SM, Thun MJ. Radon and nonrespiratory mortality in the American Cancer Society cohort. American Journal of Epidemiology. 2012;**176**(9):808-814. DOI: 10.1093/aje/kws198

[54] Turner MC, Krewski D, Chen Y, Pope CA 3rd, Gapstur SM, Thun MJ. Radon and COPD mortality in the American Cancer Society cohort. The European Respiratory Journal. 2012;**39**(5):1113-1119. DOI: 10.1183/09031936.00058211

[55] Jostes RF. Genetic, cytogenetic and carcinogenic effects of radon: A review. Mutation Research—Reviews in Genetic Toxicology. 1996;**340**:125-139. DOI: 10.1016/S0165-1110(96) 90044-5

[56] Hubaux R, Becker-Santos DD, Enfield KS, Lam S, Lam WL, Martinez VD. Arsenic, asbestos and radon: Emerging players in lung tumorigenesis. Environmental Health. 2012;**22**:11-89. DOI: 10.1186/1476-069X-11-89

[57] Tirmarche M, Harrison JD, Laurier D, Paquet F, Blanchardon E, Marsh JW. International Commission on Radiological Protection. ICRP Publication 115. Lung cancer risk from radon and progeny and statement on radon. Annals of the ICRP. 2010;**40**:1-64. DOI: 10.1016/j.icrp.2011.08.011

[58] M (PI) F, Abelenzev V, Bensman V. Studies on the negative impact of natural radioactivity (radon) on public health (final). In: Ministry of Environmental Protection of the Republic of Kazakhstan. "Ecoservice-C". Final Report. Almaty; 2007 233 p. (in Russian)

[59] Apsalikov K, Lipikhina A, Kolbin D, Mansarina A, Aleksandrova L, Brait Y, Zhakupova S. Radon on the territory of Kalachi village in Akmola region. Universum: Chemistry and Biology. 2015;**12**(19) (in Russian)

[60] Bersimbaev RI. Chromosomal analysis of mutagenesis level on uranium mining/millig industry workers in the Northern Kazakhstan. Vestnik Kaz Nat Univ. 2002;**2**(17) (in Russian)

[61] Hellman B, Friis L, Vaghef H, Edling C. Alkaline single cell gel electrophoresis and human biomonitoring for genotoxicity: A study on subjects with residential exposure to radon. Mutation Research. 1999;**442**:121-132. DOI: 10.1016/S1383-5718(99)00083-2

[62] Li BY, Tong J. Adverse effects attributed to long-term radon inhalation in rats. Journal of Toxicology and Environmental Health. Part A. 2007;**70**(11):925-930. DOI: 10.1080/15287390701290162

[63] Abo-Elmagd M, Daif MM, Eissa HM. Cytogenetic effects of radon inhalation. Radiation Measurements. 2008;**43**(7):1265-1269. DOI: 10.1016/j.radmeas.2008.02.010

[64] Druzhinin VG, Sinitsky MY, Larionov AV, Volobaev VP, Minina V, Golovina TA. Assessing the level of chromosome aberrations in peripheral blood lymphocytes in long-term resident children under conditions of high exposure to radon and its decay products. Mutagenesis. 2015;**30**(5):677-683. DOI: 10.1093/mutage/gev029

[65] Meenakshi C, Mohankumar MN. Synergistic effect of radon in blood cells of smokers—An in vitro study. Mutation Research. 2013;**757**(1):79-82. DOI: 10.1016/j.mrgentox.2013.06.018

[66] Zölzer F, Havránková R, Freitinger Skalická Z, Rössnerová A, Šrám RJ. Analysis of genetic damage in lymphocytes of former uranium processing workers. Cytogenetic and Genome Research. 2015;**147**(1):17-23. DOI: 10.1159/000441889

[67] Vähäkangas KH, Samet JM, Metcalf RA, Welsh JA, Bennett WP, Lane DP, Harris CC. Mutations of p53 and ras genes in radon-associated lung cancer from uranium miners. Lancet. 1992;**339**:576-580. DOI: 10.1016/0140-6736(92)90866-2

[68] Taylor JA, Watson MA, Devereux TR, Michels RY, Saccomanno G, Anderson M. p53 mutation hotspot in radon associated lung cancer. Lancet. 1994;**343**:86-87. DOI: 10.1016/S0140-6736(94)90818-4

[69] Popp W, Vahrenholz C, Schuster H, Wiesner B, Bauer P, Täuscher F, Plogmann H, Morgenroth K, Konietzko N, Norpoth K. p53 mutations and codon 213 polymorphism of p53 in lung cancers of former uranium miners. Journal of Cancer Research and Clinical Oncology. 1999;**125**(5):309-312

[70] McDonald JW et al. p53 and K-ras in radon-associated lung adenocarcinoma. Cancer Epidemiology, Biomarkers & Prevention. 1995;**4**:791-793

[71] Yngveson A, Williams C, Hjerpe A, Lundeberg J, Söderkvist P, Pershagen G. p53 mutations in lung cancer associated with residential radon exposure. Cancer Epidemiology, Biomarkers & Prevention. 1999;**8**:433-438

[72] Lo YM, Darby S, Noakes L, Whitley E, Silcocks PB, Fleming KA, Bell JI. Screening for codon 249 p53 mutation in lung cancer associated with domestic radon exposure. Lancet. 1995;**345**:60. DOI: 10.1016/S0140-6736(95)91183-9

[73] Hollstein M, Bartsch H, Wesch H, Kure EH, Mustonen R, Mühlbauer KR, Spiethoff A, Wegener K, Wiethege T, Müller KM. p53 gene mutation analysis in tumors of patients exposed to alpha-particles. Carcinogenesis. 1997;**18**(3):511-516

[74] Chang TC, Wentzel EA, Kent OA, Ramachandran K, Mullendore M, Lee KH, Feldmann G, Yamakuchi M, Ferlito M, Lowenstein CJ, Arking DE, Beer MA, Maitra A, Mendell JT. Transactivation of miR-34a by p53 broadly influences gene expression and promotes apoptosis. Molecular Cell. 2007;**26**:745-752. DOI: 10.1016/j.molcel.2007.05.010

[75] He L, He X, Lim LP, de Stanchina E, Xuan Z, Liang Y, Xue W, Zender L, Magnus J, Ridzon D, Jackson AL, Linsley PS, Chen C, Lowe SW, Cleary MA, Hannon GJ. A microRNA component of the p53 tumour suppressor network. Nature. 2007;**447**:1130-1134. DOI: 10.1038/nature05939

[76] Corney DC, Flesken-Nikitin A, Godwin AK, Wang W, Nikitin AY. MicroRNA-34b and microRNA-34c are targets of p53 and cooperate in control of cell proliferation and adhesion-independent growth. Cancer Research. 2007;**67**:8433-8438. DOI: 10.1158/0008-5472.CAN-07-1585

[77] Jiang L, Huang Q, Chang J, Wang E, Qiu X. MicroRNA HSA-miR-125a-5p induces apoptosis by activating p53 in lung cancer cells. Experimental Lung Research. 2011;**37**:387-398. DOI: 10.3109/01902148.2010.492068

[78] Suzuki HI, Yamagata K, Sugimoto K, Iwamoto T, Kato S, Miyazono K. Modulation of microRNA processing by p53. Nature. 2009;**460**:529-533. DOI: 10.1038/nature08199

[79] Valencia-Quintana R, Sánchez-Alarcón J, Tenorio-Arvide MG, Deng Y, Montiel-González JM, Gómez-Arroyo S, Villalobos-Pietrini R, Cortés-Eslava J, Flores-Márquez AR, Arenas-Huertero F. The microRNAs as potential biomarkers for predicting the onset of aflatoxin exposure in human beings: A review. Frontiers in Microbiology. 2014;**5**:1-14. DOI: 10.3389/fmicb.2014.00102

[80] Izzotti A, Pulliero A. The effects of environmental chemical carcinogens on the microRNA machinery. International Journal of Hygiene and Environmental Health. 2014;**217**(6):601-627. DOI: 10.1016/j.ijheh.2014.01.001

[81] Jayanthy A, Setaluri V. Light-regulated MicroRNAs. Photochemistry and Photobiology. 2015;**91**:163-172. DOI: 10.1111/php.12386

[82] Metheetrairut C, Slack FJ. MicroRNAs in the ionizing radiation response and in radiotherapy. Current Opinion in Genetics & Development. 2013;**23**:12-19. DOI: 10.1016/j.gde.2013.01.002

[83] Czochor JR, Glazer PM. microRNAs in cancer cell response to ionizing radiation. Antioxidants & Redox Signaling. 2014;**21**:293-312. DOI: 10.1089/ars.2013.5718

[84] Halimi M, Asghari SM, Sariri R, Moslemi D, Parsian H. Cellular response to ionizing radiation: A microRNA story. International Journal of Molecular and Cellular Medicine. 2012;**1**:178-184

[85] Izzotti A, Calin GA, Steele VE, Croce CM, De Flora S. Relationship of microRNA expression in lung with age and exposure to cigarette smoke and light. The FASEB Journal. 2009;**23**:3243-3250. DOI: 10.1096/fj.09-135251

[86] Chaudhry MA. Radiation-induced microRNA: Discovery, functional analysis, and cancer radiotherapy. Journal of Cellular Biochemistry. 2014;**115**:436-449. DOI: 10.1002/jcb.24694

[87] Simone NL, Soule BP, Ly D, Saleh AD, Savage JE, Degraff W, Cook J, Harris CC, Gius D, Mitchell JB. Ionizing radiation-induced oxidative stress alters miRNA expression. PLoS One. 2009;**4**(7):e6377. DOI: 10.1371/journal.pone.0006377

[88] Shin S, Cha HJ, Lee EM, Lee SJ, Seo SK, Jin HO, Park IC, Jin YW, An S. Alteration of miRNA profiles by ionizing radiation in A549 human non-small cell lung cancer cells. International Journal of Oncology. 2009;**35**:81-86. DOI: 10.3892/ijo_00000315

[89] Chaudhry MA. Real-time PCR analysis of micro-RNA expression in ionizing radiation-treated cells. Cancer Biotherapy & Radiopharmaceuticals. 2009;**24**:49-56. DOI: 10.1089/cbr.2008.0513

[90] Templin T, Young EF, Smilenov LB. Whole mouse blood microRNA as biomarkers for exposure to -rays and (56) Fe ion. International Journal of Radiation Biology. 2011;**87**:653-662. DOI: 10.3109/09553002.2012.690549

[91] Templin T, Young EF, Smilenov LB. Proton radiation-induced miRNA signatures in mouse blood: Characterization and comparison with 56Fe-ion and gamma radiation. International Journal of Radiation Biology. 2012;**88**:531-539. DOI: 10.3109/09553002.2012.690549

[92] Templin T, Paul S, Amundson SA, Young EF, Barker CA, Wolden SL, Smilenov LB. Radiation induced micro-RNA expression changes in peripheral blood cells of radiotherapy patients. International Journal of Radiation Oncology, Biology, Physics. 2011;**80**:549-557. DOI: 10.1016/j.ijrobp.2010.12.061

[93] Tufman A, Tian F, Huber RM. Can microRNAs improve the management of lung cancer patients? A clinician's perspective. Theranostics. 2013;**3**:953-963. DOI: 10.7150/thno.6615

[94] Boeri M, Pastorino U, Sozzi G. Role of microRNAs in lung cancer: microRNA signatures in cancer prognosis. Cancer Journal. 2012;**18**(3):268-274. DOI: 10.1097/PPO.0b013e318258b743

[95] Sozzi G, Boeri M, Rossi M, Verri C, Suatoni P, Bravi F, Roz L, Conte D, Grassi M, Sverzellati N, Marchiano A, Negri E, La Vecchia C, Pastorino U. Clinical utility of a plasma-based miRNA signature classifier within computed tomography lung cancer screening: A correlative MILD trial study. Journal of Clinical Oncology. 2014;**32**(8):768-773. DOI: 10.1200/JCO.2013.50.4357

[96] Wang Y, Scheiber MN, Neumann C, Calin GA, Zhou D. MicroRNA regulation of ionizing radiation-induced premature senescence. International Journal of Radiation Oncology, Biology, Physics. 2011;**81**:839-848. DOI: 10.1016/j.ijrobp.2010.09.048

[97] Ma W, Yu J, Qi X, Liang L, Zhang Y, Ding Y, Lin X, Li G, Ding Y. Radiation-induced microRNA-622 causes radioresistance in colorectal cancer cells by down-regulating Rb. Oncotarget. 2015;**6**:15984-15994. DOI: 10.18632/oncotarget.3762

[98] Liu S, Pan X, Yang Q, Wen L, Jiang Y, Zhao Y, Li G. MicroRNA-18a enhances the radiosensitivity of cervical cancer cells by promoting radiation-induced apoptosis. Oncology Reports. 2015;**33**(6):2853-2862. DOI: 10.3892/or.2015.3929

[99] Lan F, Yue X, Ren G, Li H, Ping L, Wang Y, Xia T. miR-15a/16 enhances radiation sensitivity of non-small cell lung cancer cells by targeting the TLR1/NF-κB signaling pathway. International Journal of Radiation Oncology, Biology, Physics. 2015;**91**(1):73-81. DOI: 10.1016/j.ijrobp.2014.09.021

[100] Joly-Tonetti N, Viñuelas J, Gandrillon O, Lamartine J. Differential miRNA expression profiles in proliferating or differentiated keratinocytes in response to gamma irradiation. BMC Genomics. 2013;**14**:184. DOI: 10.1186/1471-2164-14-184

[101] Lee KF, Chen YC, Hsu PWC, Liu IY, Wu LSH. MicroRNA Expression Profiling Altered by Variant Dosage of Radiation Exposure. BioMed Research International. 2014;**2014**:456323. DOI: 10.1155/2014/456323

[102] Naito Y, Oue N, Pham TT, Yamamoto M, Fujihara M, Ishida T, Mukai S, Sentani K, Sakamoto N, Hida E, Sasaki H, Yasui W. Characteristic miR-24 expression in gastric cancers among atomic bomb survivors. Pathobiology. 2015;**82**(2):68-75. DOI: 10.1159/000398809

[103] Chen Z, Wang D, Gu C, Liu X, Pei W, Li J, Cao Y, Jiao Y, Tong J, Nie J. Down-regulation of let-7 microRNA increased K-ras expression in lung damage induced by radon.

Environmental Toxicology and Pharmacology. 2015;**40**(2):541-548. DOI: 10.1016/j.etap.2015.08.009

[104] Cui FM, Li JX, Chen Q, HB D, Zhang SY, Nie JH, Cao JP, Zhou PK, Hei TK, Tong J. Radon-induced alterations in micro-RNA expression profiles in transformed BEAS2B cells. Journal of Toxicology and Environmental Health. Part A. 2013;**76**(2):107-119. DOI: 10.1080/15287394.2013.738176

[105] Pei WW, Tao LJ, Zhang LSW, Zhang SY, Cao JP, Jiao Y, Tong J, Nie JH, Circular RNA. Profiles in mouse lung tissue induced by radon. Environmental Health and Preventive Medicine. 2017;**22**:36. DOI: 10.1186/s12199-017-0627-6

[106] Robertson A, Allen J, Laney R, Curnow A. The cellular and molecular carcinogenic effects of radon exposure: A review. International Journal of Molecular Sciences. 2013;**14**:14024-14063. DOI: 10.3390/ijms140714024

[107] Masoomi JR, Mohammadi S, Amino M, Ghiassi-Nejad M. High background radiation areas of Ramsar in Iran: Evaluation of DNA damage by alkaline single cell gel electrophoresis (SCGE). Journal of Environmental Radioactivity. 2006;**86**:176-186. DOI: 10.1016/j.jenvrad.2005.08.005

[108] Fernandez-Capetillo O, Lee A, Nussenzweig M, Nussenzweig A. H2AX: The histone guardian of the genome. DNA Repair. 2004;**3**:959-067. DOI: 10.1016/j.dnarep.2004.03.024

[109] Scott BR, Belinsky SA, Leng S, Lin Y, Wilder JA, Damiani LA. Radiation-stimulated epigenetic reprogramming of adaptive-response genes in the lung: An evolutionary gift for mounting adaptive protection against lung cancer. Dose Response. 2009;**7**(2):104-131. DOI: 10.2203/dose-response.08-016.Scott

[110] Herman JG, Baylin SB. Gene silencing in cancer in association with promoter hypermethylation. The New England Journal of Medicine. 2003;**349**:2042-2054. DOI: 10.1056/NEJMra023075

[111] Su S, Jin Y, Zhang W, Yang L, Shen Y, Cao Y, Tong J. Aberrant promoter methylation of p16(INK4a) and O(6)-methylguanine-DNA methyltransferase genes in workers at a Chinese uranium mine. Journal of Occupational Health. 2006;**48**(4):261-266. DOI: 10.1539/joh.48.261

[112] Wang BH, Li YY, Han JZ, Zhou LY, Lv YQ, Zhang HL, Zhao L. Gene methylation as a powerful biomarker for detection and screening of non-small cell lung cancer in blood. Oncotarget. 2017;**8**(19):31692-31704. DOI: 10.18632/oncotarget.15919

[113] Yang Z, Li F. O-6-methylguanine-DNA methyltransferase gene promoter methylation and lung cancer risk: A meta-analysis. Journal of Cancer Research and Therapeutics. 2016;**12**(Supplement):233-236. DOI: 10.4103/0973-1482.200745

[114] Belinsky SA, Klinge DM, Liechty KC, March TH, Kang T, Gilliland FD, Sotnic N, Adamova G, Rusinova G, Telnov V. Plutonium targets the p16 gene for inactivation by promoter hypermethylation in human lung adenocarcinoma. Carcinogenesis. 2004;**25**(6):1063-1067. DOI: 10.1093/carcin/bgh096

[115] Toyooka S, Suzuki M, Tsuda T, Toyooka KO, Maruyama R, Tsukuda K, Fukuyama Y, Iizasa T, Fujisawa T, Shimizu N, Minna JD, Gazdar AF. Dose effect of smoking on aberrant methylation in nonsmall cell lung cancers. International Journal of Cancer. 2004;**110**:462-464. DOI: 10.1002/ijc.20125

[116] Divine KK, Pulling LC, Marron-Terada PG, et al. Multiplicity of abnormal promoter methylation in lung adenocarcinomas from smokers and never smokers. International Journal of Cancer. 2005;**114**:400-405. DOI: 10.1002/ijc.20761

[117] Bastide K, Guilly MN, Bernaudin JF, Joubert C, Lectard B, Levalois C, Malfoy B, Chevillard S. Molecular analysis of the Ink4a/Rb1-Arf/Tp53 pathways in radon-induced rat lung tumors. Lung Cancer. 2009;**63**(3):348-353. DOI: 10.1016/j.lungcan.2008.06.007

[118] Pulling LC, Divine KK, Klinge DM, Gilliland FD, Kang T, Schwartz AG, Bocklage TJ, Belinsky SA. Promoter hypermethylation of the O6-methylguanine-DNA methyltransferase gene: More common in lung adenocarcinomas from never-smokers than smokers and associated with tumor progression. Cancer Research. 2003;**63**(16):4842-4848

[119] USEPA (U.S. Environmental Protection Agency) Home Buyers and Sellers Guide to Radon. 1993. Report 402-R-93-003. [Internet]. 2006. Available at http://www.radon-levels.com [Accessed: 2017-09-15]

[120] Scott BR, Sanders CL, Mitchel REJ, Boreham DR. CT scans may reduce rather than increase the risk of cancer. Journal of the American Physicians and Surgeons. 2008;**13**(1):8-11

[121] Thompson RE, Nelson DF, Popkin JH, Popkin A. Case-control study of lung cancer risk from residential radon exposure in Worchester County, Massachusetts. Health Physics. 2008;**94**(3):228-241. DOI: 10.1097/01.HP.0000288561.53790.5f

[122] Pogribny I, Koturbash I, Tryndyak V, Hudson D, Stevenson SM, Sedelnikova O, et al. Fractionated low-dose radiation exposure leads to accumulation of DNA damage and profound alterations in DNA and histone methylation in the murine thymus. Molecular Cancer Research. 2005;**3**:553-561. DOI: 10.1158/1541-7786.MCR-05-0074

[123] http://www.worldlifeexpectancy.com [Accessed: 2017-09-15]

[124] Sagyndykova G. Morphological and molecular biological features of lung cancer that developed under conditions of increased radiation [thesis]. Moscow Medical Academy: Moscow; 2002 (in Russian)

[125] Dubrova YE, Bersimbaev RI, Djansugurova LB, Tankimanova MK, Mamyrbaeva ZZ, Mustonen R, Lindholm C, Hultén M, Salomaa S. Nuclear weapons tests and human germline mutation rate. Science. 2002;**295**(5557):1037. DOI: 10.1126/science.1068102

[126] Vasil'eva ZZ, Bersimbaev RI, Bekmanov BO, Vorobtsova IE. The polymorphism of DNA repair genes XRCC1, XRCC3 and the level of chromosomal aberrations in the uranium workers. Radiatsionnaia Biologiia, Radioecologiia. 2012;**52**(1):25-30 (in Russian)

[127] Larionov AV, Sinitsky MY, Druzhinin VG, Volobaev VP, Minina VI, Asanov MA, Meyer AV, Tolochko TA, Kalyuzhnaya EE. DNA excision repair and double-strand break repair

gene polymorphisms and the level of chromosome aberration in children with long-term exposure to radon. International Journal of Radiation Biology. 2016;**92**(8):466-474. DOI: 10.1080/09553002.2016.1186303

[128] Izzotti A, Carozzo S, Pulliero A, Zhabayeva D, Ravetti JL, Bersimbaev R. Extracellular MicroRNA in liquid biopsy: Applicability in cancer diagnosis and prevention. American Journal of Cancer Research. 2016;**6**(7):1461-1493

[129] Geretto M, Pulliero A, Rosano C, Zhabayeva D, Bersimbaev R, Izzotti A. Resistance to cancer chemotherapeutic drugs is determined by pivotal microRNA regulators. American Journal of Cancer Research. 2017;**7**(6):1350-1371

Chapter 8

Radon Calibration System

Ayman M. Abdalla, Ahmed M. Ismail and
Tayseer I. Al-Naggar

Additional information is available at the end of the chapter

http://dx.doi.org/10.5772/intechopen.69900

Abstract

In this chapter, an irradiation radon system was explained in detail. This system is based on soil gas as a natural radon source. The radon system was used to determine both the calibration factor of the radon detector and the equilibrium factor between radon and its short-lived daughters. Also, the calibration factor has been calculated theoretically and experimentally. The effect of humidity upon the calibration factor has been investigated. The diffusion of radon through polyethylene membrane has been determined using new nuclear method. This method depends upon the physical decay of radon.

Keywords: radon detection techniques, calibration system

1. Introduction

Radon $^{222}_{86}Rn$ is a naturally occurring noble element, radioactive gas, colorless, odorless, soluble in water, and chemically inert radioactive element. The importance of radon gas comes from its benefits in the field of uranium exploration, testing the permeability of gases in polymer membranes and its harmful effects on health. In this chapter, we will try to discuss briefly radon sources, methods of radon detection, and radon calibration system including the methods of determining the calibration factor for radon measurement experimentally and theoretically. The measurement of radon permeability through polyethylene membranes using scintillation detectors will also be discussed as a useful application in the field of radon measurement.

2. Radon sources

The three naturally produced isotopes, radon (^{222}Rn from ^{238}U series), thoron or thorium emanation (^{220}Rn from ^{232}Th series), and action or actinium emanation (^{219}Rn from ^{235}U series), are all disintegrated by producing alpha particles (**Table 1**) [1].

Isotope	Common name	Half life	Decay series
^{222}Rn	Radon	3.8 days	^{238}U
^{220}Rn	Thoron	54.5 s	^{232}Th
^{219}Rn	Actinon	3.92 s	^{235}U

Table 1. Radon isotopes.

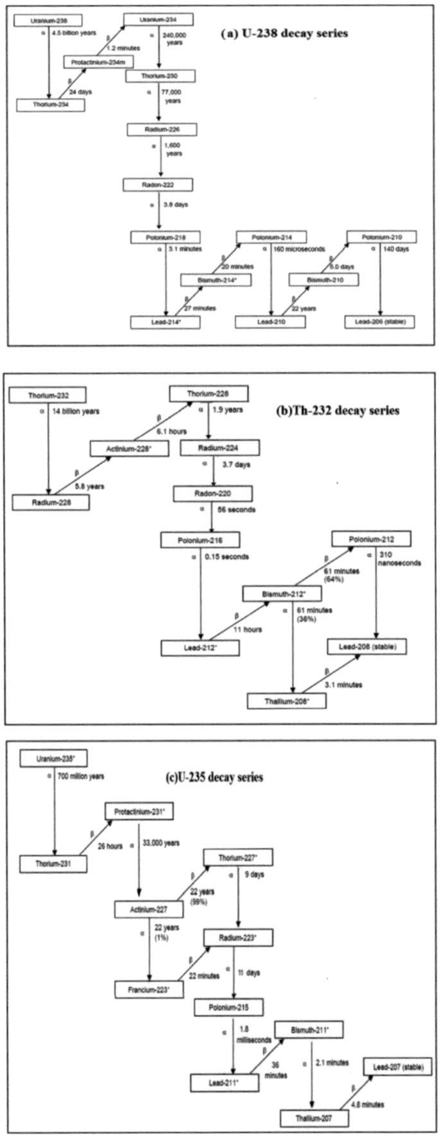

Figure 1. Three natural decay series [3].

Radon produced from uranium decay series. It is known that all types of rocks contain various amounts of uranium especially mountainous areas [2]. Radium is produced through radioactive decay of the three natural chains (^{238}U, ^{235}U, or ^{232}Th), and radon isotopes are produced as a result of radium decay (**Figure 1a–c**).

3. Radon measurement techniques

Measurement, and detection of radon can done by two ways directly by itself (^{222}Rn only) or indirectly by its progenies. Detection and measurement of radon can be accomplished with α, β, or γ radiation detection. One of the following techniques can be used for detection: nuclear emulsion, adsorption, solid scintillation, gamma spectroscopy, beta monitoring, solid-state nuclear track detector (SSNTD), electrometer or electroscope, ionization chambers, surface barrier detectors, thermoluminescent phosphors, electret, and collection. Radon-222 can be measured by passive method when radon entering the detection volume by just diffusion, or active, which contains the pushing of gas into or through a detecting instrument.

3.1. Passive devices

The detection sensor is commonly put inside a vessel that has an opening to allow radon enter it. The vessel is intended to keep the detector and to make area about the detector for a sensitive size huge enough so as to have as many α-particles created and detected in as small time as possible, and in the case of direct measurement of the radiation, an inverted plastic or paper cup in various versions [4] or more advanced container can be also used.

3.1.1. Solid-state nuclear track detector

Alpha particle is detected in a SSNTD by a particular path or "trail" of damage, which after chemical etching produces narrow channel, and it is seen under the microscope. SSNTDs such as CR-39, cellulose esters, and polycarbonates like bis-phenol-A polycarbonate are sensitive to α-particles within the energy range of the particles produced by radon. The SSNTDs are more sensitive to charged particles, on the other side, SSNTDs are more insensitive to β and γ rays, so beta and gamma rays do not serve an etchable discrete trails or tracks. However, preirradiation with nontrack forming radiation has also been found to increase sensitivity and etching speed [5]. There are various types of polymer detectors such as polyallyldiglycol carbonate (CR-39), cellulose nitrate (LR-115), and polycarbonate (Lexan, Makrofol) [6, 7]. SSNTDs do not need an energy source to work, so their detecting property is an essential quality of the substance they are made of. They are consequently very well suitable for long-term measurements in the field, so they have received special awareness. The methods in which SSNTDs are used for the detection of radon and/or radon progenies which, change generally in the geometry of the arrangement, commonly the use or not of diffusion membranes; and the use (or not) of filters to keep progenies out of equipment [8]. In other states, the detector is located vertically [9]. Some procedures have two measuring chambers, detached or not by a filter and/or a membrane. In some cases, SSNTDs are used in conjunction with another type of detector. For instant SSNTDs can be combined with electrets. In this way, a prompt response can be obtained together with an energy differentiation [10].

3.1.2. Alpha card

The alpha card technique is a passive radon detecting process which supplies a means for accurate measurements of radon in soil gas, with dimensions 4.5 cm × 5 cm, and diameter hole (205 cm) punctured in the interior of it. The hole is completely closed with a slight film. Silicon detectors are used to measure the activity of alpha card in terms of alpha particles. The evaluation of reading method consists of two silicon detectors facing one another, between which the card is placed. A slightly good counting efficiency is attained because the assembling membrane is very thin, which alpha particles issuing from both sides of the card impinge upon either one of the silicon detector. The alpha card reader is a small, movable with operated battery supply that can be used in the field, and the common reading time is about 5 min [11].

3.1.3. Electrets detector

The electret radon monitor is a new technique which is used in the field to measure the radon concentration [12]. Electret is a portion of dielectric material that displays a nearly lasting electrical charge if not otherwise worried. This charge yields a robust electrostatic field which is able to assemble ions of the contrary sign and the overall charge of the electret reductions. This property can be used to make gamma radiation detectors away from the electrets [13]. The concept is not enough till the modern progress of high dielectric fluorocarbon polymers, which have turned electrets to be trustworthy, electronic constituents until under high temperature and moisture provisions. The main components of a radon monitor electret are steel canister and the electret dosimeter which is stable on the internal upper. At the lower of such canister little inlet allows the radon gas to reach the assembly during a filter. The ions are produced from radon which produces ionizing particles inside the volume of the device. The ions are assembled, and the full charge of the electrets changes; after some time has passed, the surface potential of the electret is measured by the shutter method [14–16] using a battery-operated, lightweight piece of equipment.

Detector sensibility and dynamic field are based on the size of the chamber of the equipment and of the thickness of the sensitive substance. As for solid-state nuclear track detector (SSNTD), varied geometrical preparations have been designed. For case, the measurement can be actuated time wise by having a dynamic segment translating, the electret in the lodging [17]. Detection of radon alone is achieved by using a 4π-geometry chamber, on the wall of which another electret is placed for daughter detection [18]. In some other instances, a positively charged electrode can be coated with a SSNTD or with a TL phosphor for completeness of detection and measurement.

3.1.4. The mechanism of chemical absorption

In chemical absorption mechanism, the radon dimension structure constructed on the chemical absorption of the gas on an adsorption medium such as charcoal, and then on the determination the activity of the restricted daughters ^{214}Pb and ^{214}Bi. When the adsorbing medium is actuated charcoal, the system is well with name ROAC (radon on activated charcoal) technique. The instrument for the chemical absorption of the radon is aplastic even when the can is placed inside the place contain radon gas, with its lid open, and left in place for a few days, typically from 4 to 12 days. The container is then returned, and the cover locked. It is then transported to the laboratory where gamma activity is specified using NaI scintillation counter. Radon activity can be achieved at the check point from the decay activity curves, after adequate calibration.

By this method, thoron assay is disappeared, and the count rate must be for radon only [19] found it is possible to determine the ratio between thoron and radon emanation from a soil. Another way of determining the amount of radon is to desorb it into scintillation liquid and to count the alpha, and beta activity by liquid scintillation [20].

3.1.5. Thermoluminescent detectors

When an ionizing particle moves through a crystal, some electrons liberate from its main sites into the conduction band. The electrons are then restricted into faults of the crystal and kept there in a very constant site, and the electron can leak if the crystal is heated. As a result of recombination with the holes in the valence band, or with recombination centers, by process of radioactive, light is liberated [21, 22]. Thermoluminescent dosimeters (TLDs) are built on this attitude, in the state of radon discovering, which depends upon the alpha activity of the radon progenies. Radon can inside a detection volume include the TLD.

At a small distance, the steel plate is placed facing TLD. Radon progenies summated on the plate and eventually decay, thus creating energy storage in the TLD. After prober exposure to the radon-rich atmosphere, the TLD is recovered and "read" in a TLD apparatus. The concentration of radon is appearing on a TLD device after suitable exposure to the radon abundant air. This device is basically a little furnace, and here the heat is advanced under a nitrogen atmosphere at a very fully controlled rate. The TLD produces quantity of light that falls on the photocathode of a photomultiplier and then is transported into an electrical signal. TLD spectra are noted. Every peak relates to a particular trap, and the amplitude and/or the area below a peak is proportional to the radiation dose. In many commercially available devices, various types of detectors containing TLD sensors have been used field measurements, and in the sphere of different applications. They are using them for detecting oil gas reservoirs [23].

The advanced detection method, which really keeps an electret detector for gamma-background gauge, has three TLDS [24]. One is a preirradiated TLD which provides information on the natural fading of the signal recorded in the two others. The signal related to the radioactivity created by the radon, which has spread to the detection bulk, and the next makes a cross measurement of the γ-field. In certain cases, the housing of the TLD sensor is made such that the full detector can resist the heat at which the light is emitted from the TLD when reading it [25].

3.1.6. Solid-state electronic detectors

Solid-state electronic detectors (SSEDs) are not widely used for radon detection because it needs considerable quantity of energy not simply obtainable in the field, and also it is slightly costly and brittle. One may mention, for instance, the Alpha meter, which the active part of the equipment is a Si(Li) detector associated with a counter unit, both operated on batteries. The counting is directly displayed on the instrument [1]. A special version of the radon devices allows water level, temperature, and conductivity measurements to be made in the meantime. The detector is especially produced for indoor measurements [26]. The measuring instrument with a hemispherical bell chamber is fitted with a solid-state alpha detector built by Ref. [27]. Several active cells for counting radon daughters use by Ref. [28]. The device based on light measurement with solid-state detectors is built by Ref. [29].

3.1.7. Ion chambers and electrometers

Apparatuses depend on the measurement of ion pairs produced through the way of α-particles in the air. The measurement can be achieved by using ion chambers and electrometers. At the first status of detection part, the amount of electrical charges can record, which have been formed without laying stable voltage. Through a second state, one assembles the electrical charges that have been produced between two electrodes in an electrical field [1].

3.2. Active devices

In active apparatuses, radon gas is either pumped from the surrounding or removed by means of a gas or liquid. In addition, the lively part for measuring radon or its progenies results is an electrical or electronic device in public. By applying a fan or a pump, many of the apparatuses up could be turned into an active apparatus. The detector where radon is propelled from the ground soil is built by Ref. [30]. The gas drift pushes through the electronic meter to the external of the instrument. Radon concentration is then determined. Direct scintillation can also be used for detecting the radon level by forcing air through the detector [31]. The other type of active detection is radon sampling or extract sampling. After sampling, the water or air or the solvent that holds radon is taken to the laboratory for detection. Sampling methods are various, but they all need one to be specially wary not to pollute the sample. Sampling can be accomplished by syringes, compressor, vacuum container, water sampling, degassing on the spot, suction action of pumps, and so on.

4. Radon irradiation chamber and its applications

Radon is a worldwide problem because inhalation of radon engages more than 50% from total natural radiation dose to the world population [3]. Radon and its short-lived decay products caused lung cancer in the USA according to the Environmental Protection Agency (EPA) [32], and the short-lived decay products of radon are responsible for most of the hazard by inhalation. The main environmental source is the soil gas emanation in dwelling areas or from building materials [33]. Solid-state nuclear track detector (CR-39) is used widely to determine the concentration of radon [34]. The calibration factor converts the track density to radon concentration and can be calculated from the relation below [35, 36]:

$$K = \frac{\rho}{c_0 t_0} \tag{1}$$

where ρ: track density (track cm^{-2}), C_0:radon concentration (Bq m^{-3}), t_0: exposure time (day).

The annual effective dose rate (ED, mSv/y) of inhaled radon and its progeny can be estimated according to the following equation [37]:

$$\text{ED} = C_0 FTD \tag{2}$$

where F: equilibrium factor, D: dose conversion factor (9 nSv h^{-1}/Bq m^{-3}), T: average indoor occupancy time per person (7000 h y^{-1}), C_0: radon activity concentration in air (Bq m^{-3}).

The equilibrium factor is defined as the ratio of potential alpha energy concentration (PAEC) of actual air radon mixture (also radon progeny) to the PAEC in secular equilibrium with radon [38].

$$F = \frac{(\text{PACE})_{\text{air radon mixture}}}{(\text{PACE})_{\text{equilibrium}}} \tag{3}$$

There are two methods to estimate the equilibrium factor: one method uses silicon surface barrier detector by the following equation [39]:

$$F = \gamma_1 \frac{C_1}{C_0} + \gamma_2 \frac{C_2}{C_0} + \gamma_3 \frac{C_3}{C_0} + \gamma_4 \frac{C_4}{C_0} \tag{4}$$

where C_0, C_1, C_2, C_3, and C_4 are the activity concentrations of ^{222}Rn and its short-lived progenies ^{218}Po, ^{214}Po, ^{214}Bi, and ^{214}Po, respectively, and the coefficients γ_i are constant.

In another method, a solid-state nuclear track detector was used to evaluate the equilibrium factor by the following equations [40, 41]:

$$D = K(C_0 + C_1 + C_4) \tag{5}$$

$$D_0 = K(C_0) \tag{6}$$

Eqs. (5) and (6) are related the respective track densities D and D_0 to the activity concentrations of radon and its progeny in air, and K is the detector sensitivity coefficient.

The equilibrium factor can be calculated by using the LR-115 detector, using the following equation [40]:

$$F = 0.47\left(\frac{D}{D_0}\right) - 0.47 \tag{7}$$

For the CR-39 track detector, the equilibrium factor can be determined by the following equation [42]:

$$F = ae^{\left[-b\left(\frac{D_0}{D}\right)\right]} \quad (8)$$

where the values of two constants *a, b* are 14.958 and 7.436, respectively.

4.1. Preparing a radon source

The amount of ^{226}Ra and ^{222}Rn produced in rocks and soils depends on its ^{238}U contents, and many studies considered a variable alternative of using soil gas as a radon source in the experimental studies [43]. High-performance gamma-ray spectrometry (DSPECjr2.0) was used to measure the activity of ^{226}Ra in the soil sample; 0.424 kg of soil sample was fitted into a Marinelli beaker of 250 cm^3 volume. The gamma detector has resolution 2.2 KeV and relative efficiency 55% at 1.33 MeV of ^{60}Co. The recommended operating bias negative 4500 V. ^{176}Lu testing adapter and ^{137}Cs were used for energy calibration.

The specific activity of radium was measured using the following equations [44, 45]:

$$A_s = \frac{N}{(E \times P_r \times M)} \quad (9)$$

where *N*: counting rate for a specific gamma line, *E*: detector efficiency, P_r: absolute transition probability of γ-decay, *M*: mass of the sample (kg). In this work, the average specific activity of ^{226}Ra in the soil sample was 182 ± 13 Bq kg^{-1}. Metallic cylindrical drum with 24.7 l contains different amount of soil from the environment, used as radon source, and was covered with a polyethylene membrane to make the thoron (^{220}Rn) gas outside the system. The concentration of radon was changed using different amounts of soil.

4.2. Radon irradiation system

A radon system consists of the following parts as shown in **Figure 2**:

1. 1—plastic vessel, 2—monitor, 3—lucas cell, 3, 4—printer, 5—radon source, 6, 7—PID controlled furnaces, 8, 9, 13, 14—inlet and outlet tubes, 10—internal pump, 11, 12—valves, 15—desiccation column, 16, 17—pored filter of polyethylene, 18, 19—circular port, 20—inactive dosimeter, 21—flowmeter, 22—computer interface.

2. The lucas cell 300A offers a proper tool for detection of radon. The cell was ready at a great-efficiency scintillator of silver-activated zinc sulfide. Lucas cell volume is nearly 272 ml, sensitivity of 0.037 cpm/Bq/m^3, MDA of 27.4 Bq m^{-3}, and $E\alpha$ of 4.5–9 MeV [46]. The pump of AB-5's can be used to withdrawal air into the Lucas cell by the porous polyethylene filter, and any radon gas in the spread air decays into its daughter products, some of which are liberating α-particles. Alpha particles were hit the scintillator, which lines the inner of the Lucas cell, and create light pulses which they are amplified and 30 counted in the AB-5. A microprocessor in the AB-5 is used to convert the signal into an average count value per minute (*C*), and the net count value (*N*) is obtained by correcting the background [47].

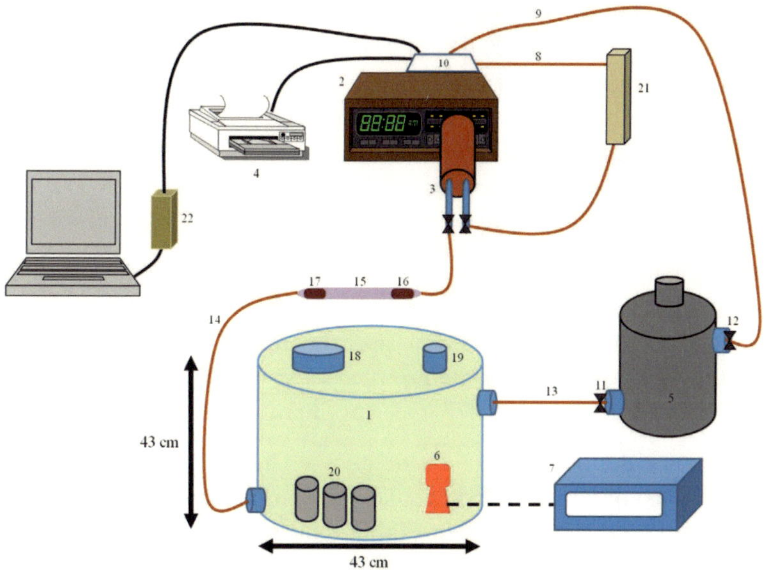

Figure 2. Radon irradiation system.

Radon gas concentration was determined by the general formula used by the AB-5 in the radon mode:

$$C_0 = \frac{NCPM}{S} \tag{10}$$

where, C_0: radon gas concentration (Bq m^{-3}), *NCPM*: the net count per minute, *S*: the counting sensitivity value (cpm/Bq/m^3).

An electronic thermometer with a liquid crystal display was used to measure the temperature inside the chamber.

4.3. Calibration factor assessment

The calibration factor of the CR-39 detector can be assessed by three cups that were used with CR-39 sheets, which had an area of 1.0 cm × 1.0 cm and a thickness of 250 μm. The plastic cups 8 cm in height and 7 cm in diameter were used in this work, and then, the detectors were etched in 6.25 N NaOH solutions at 65°C for 6 h. The etched CR-39 samples were washed in running water and dried in air. The tracks were analyzed using an optical microscope at 500×, and also the background tracks were registered.

4.4. Comparative studies with the track detector

Films from LR-115 have 12 μm thickness, which were mounted on a 100-μm transparent polyester foil, were used for two different setups was used in this work. The filtered setup is the first one, where a piece of LR-115 was stable at the end of a plastic cup. Two cups were placed in the radon irradiation system for 1 week. The exposed LR-115 sheets were etched in 2.5 N NaOH solutions at 60°C for 100 min [48]. The track density in each sheet was calculated to obtain D and D_0 for the open and closed plastic cups, respectively. The identical procedure was repeated using the CR-39 track detector that was etched at the optimum etching condition [49]. **Figure 3a** and **b** show a typical optical image of alpha particle tracks in the LR-115 and CR-39 detectors. The equilibrium factor was measured using Eqs. (7) and (8) for the LR-115 and CR-39 detectors, respectively.

Figure 4 shows the rate of track density as a function of the radon activity concentration for the diffused CR-39 track detector. From the slope of the line, it is found that the calibration factor is 0.165 track cm^2/Bq m^3 d with uncertainty approximately 20%. The obtained experimental value is considerably consistent with the reported experimental values in Farid (0.199 track cm^2/Bq m^3 d) [50] and Mansy et al. (0.177 track cm^2/Bq m^3 d) [51].

Table 2 demonstrates the equilibrium factor between radon and its progeny, which was determined using three different methods: solid-state nuclear track technique, surface barrier method, and our new radon chamber. From **Table 2**, the results obtained using our new radon chambers are considerably consistent with the results of the surface barrier detector method [7]. We can obtain a stable equilibrium factor between radon and its daughters in a closed environment of a radon system which emulates a closed environment of a room.

As shown in **Figure 5a**, the equilibrium factor tends to decrease as the temperature increases (temperature range of 20–50°C for 24 h) and this is due to the temperature impact on the equilibrium factor by two chief causes: (1) the temperature reduces the equilibrium factor by effects on the radon emanation rate and (2) the temperature rises the equilibrium factor by affecting the number of radon daughters in the air [52, 53]. Also **Figure 5b** shows that the equilibrium factor at 20°C is equivalent to 2.2 times the coefficient of equilibrium at 50°C and 1.5 times the coefficient of equilibrium at 32°C. Also, the influence of the flow rate through irradiation on the equilibrium factor between radon and its daughters was obvious. The

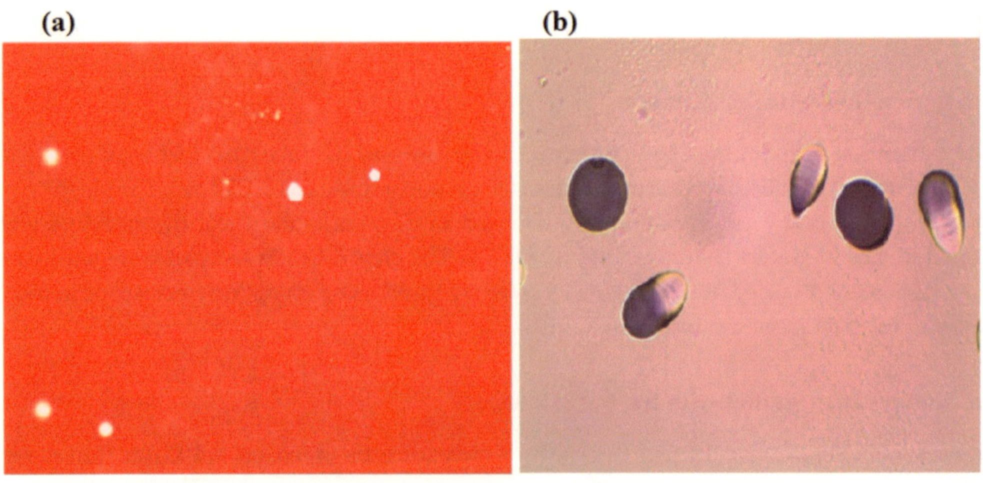

Figure 3. Track optical image of alpha particles for (a) LR-115 detector and (b) CR-39 detector.

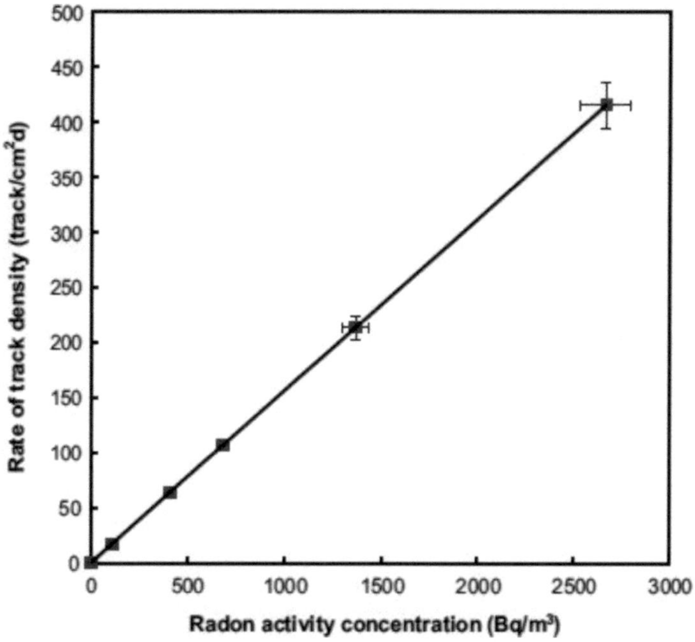

Figure 4. Rate of track density as a function of radon activity concentration.

Method	Our new radon system	SSNTD	Surface barrier detector method
Equilibrium factor	0.21	0.17 (LR-115) 0.20 (CR-39)	0.19 [5]

Table 2. Equilibriums factor obtained using three different methods.

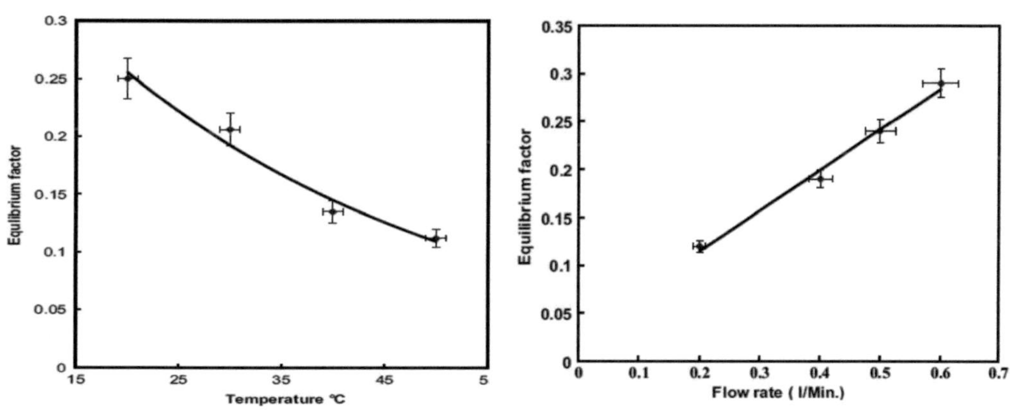

Figure 5. Equilibrium factor as a function of (a) temperature and (b) flow rate of air in the lucas cell.

system was utilized in the flow rate of 0.2–0.6 l/min for 24 h at stable heat. **Figure 5b** presents the equilibrium factor for radon, and its daughters are influenced by the variation in flow rate of radon-bearing air. The equilibrium factor rises with rising flow rate. The value of equilibrium gives 0.25 at the optimum flow rate of 0.5 l/min. These consequences indicate that the measurement of an apparent equilibrium factor as a function of the temperature and flow rate is important for determining the annual effective dose rate.

5. Radon permeability measurement through polyethylene membrane using scintillation detector

Activated charcoal technique was used to study the permeability of ^{222}Rn through polyethylene membranes. The permeability constant of ^{222}Rn through low-density polyethylene, linear low-density polyethylene, and high-density polyethylene samples has been measured, and the diffusion coefficient of radon in charcoal as well as solubility of radon in polyethylene membrane has been taken into consideration. The two gases ^{222}Rn (^{238}U-series) and ^{220}Rn (^{232}Th-series) have differences in their radioactive half-lives 3.82 days and 56 s, respectively, which provide suitable properties for their separation using proper membranes. The difference in the decay rates leads to difference in the mean diffusion distances [54, 55].

The permeability (K) of a polymer to a gas can be defined by

$$K = SD \tag{11}$$

where S: the solubility, D: the diffusion coefficient of the gas in the polymer.

The significant factor which determines the appropriateness of a membrane for separation radon isotopes is its permeability constant. Ramachandran et al. [56] suggested a method to determine the permeability constant that depends on SSNTD technique, as shown in **Figure 6**, and it is expressed by the following relation:

$$R = \frac{n_2}{n_1} = \frac{KA(V_1 + V_2)}{KA(V_1 + V_2) + V_1 V_2 \lambda \delta} \tag{12}$$

where R: ratio between the concentrations of gaseous in volume V_2 with a membrane and the concentration without any membrane in V_1, K: the permeability constant (cm^2 s^{-1}), d: the thickness of the membrane (in cm), A: the surface area of the membrane (cm^2), and λ: the decay constant (s^{-1}).

Figure 6. Schematic diagram shows solid-state nuclear track technique.

http://dx.doi.org/10.5772/intechopen.69900

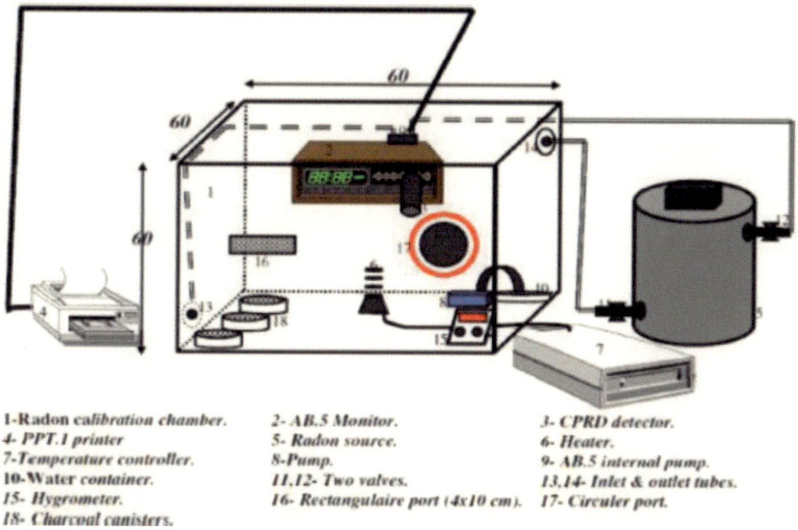

Figure 7. A schematic for radon calibration chamber.

The process used is based on exposing activated charcoal canister to radon concentration inside radon chamber [57], as illustrated in **Figure 7**.

The radon activity adsorbed by a diffusion barrier charcoal canister (DBCC) is described by the following relations given by Refs. [58–60]:

$$Q_i = \beta m C_i = V C_i \tag{13}$$

$$\frac{dC_i}{dt} = -\left(\frac{KA}{\delta V} + \lambda\right) C_i + \frac{KA}{\delta V} C_0 \tag{14}$$

$$\tau = \left(\frac{KA}{\delta V} + \lambda\right) \text{ if } P = \frac{KA}{\delta} \text{ and } \psi = \frac{1}{\tau}, \text{ Then } \psi = \frac{P}{V} + \lambda \tag{15}$$

where Q_i: the activity of radon that passes through the membrane and adsorbed by DBCC (Bq), V: equivalent to a volume of air available to the activated charcoal (cm^3), b: the adsorption coefficient of radon by activated charcoal (Bq g^{-1} per Bq cm^3), m: the mass of activated charcoal in the canister (g), C_i: the radon concentration within the canister (Bq m^{-3}), C_0: is the radon concentration in the environment air (Bq m^{-3}), Λ: the decay constant of radon (s^{-1}), K: the radon permeability (cm^2 s^{-1}),, A: cross section of the diffusion barrier (cm^2), D: thickness of the diffusion barrier (cm), and S: is the integration time constant of radon in the DBCC (s).

For $C_i = C_{i0}$, $C_0 = 0$ at $t = 0$, the solution of the Eq. (14) is

$$C_i = C_{i0} \exp\left(\frac{t}{\tau}\right) \tag{16}$$

The radon permeability for radon in the tested membrane can be easily evaluated by using the following equation:

$$K = \frac{VL}{A}(\psi - \lambda) \tag{17}$$

Permeability of radon was measured at room temperature and a comparative moisture of 20% at exposure was finished. The canisters were closed and left for 3 h. Gamma spectroscopy 3″ × 3″ NaI (TI) was used to determine the gamma spectrum from each canister, as illustrated in **Figure 8**. The range of energy for gamma lines between 295 and 609 keV was achieved, and the activity of radon in the canister was determined.

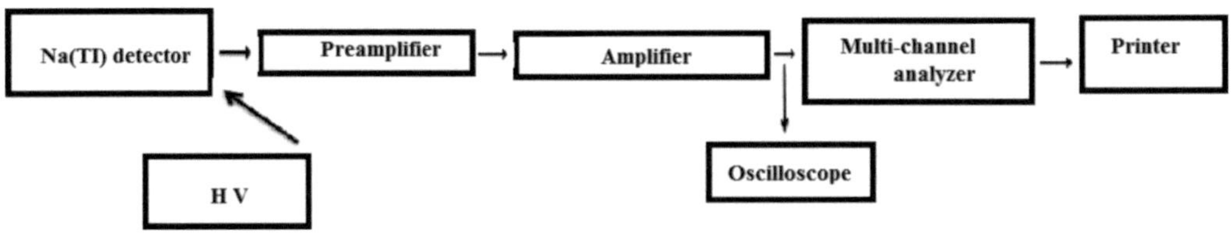

Figure 8. Schematic of the scintillation spectrometer.

Figure 9 presents the difference of the relative net count per minute as a function of time for different samples.

For SSND, pieces of CR-39 track detectors of 250 mm (TASTRAK supplied by Bristol, UK) were used for two diverse setups. In the first setup (named: filtered setup), a piece of CR-39 was stabled in the end of conical flask (250 ml), and then, the conical flask was closed with a

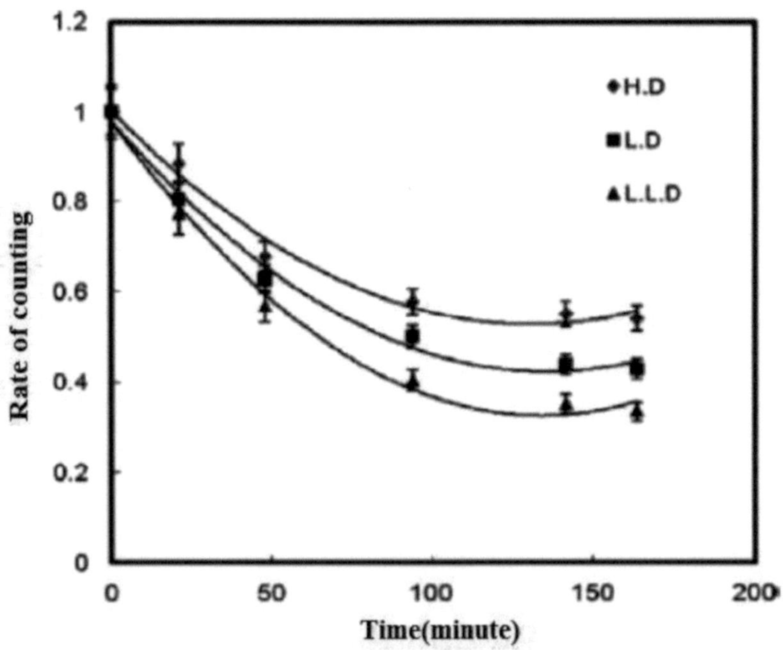

Figure 9. Rate of counting as a function of time for different polyethylene membranes.

polyethylene membrane which acts as a filter. Secondly, named bare (opened) setup, in which a piece of CR-39 was stabled at the end of opened conical flask. Filtered and open conical flasks were put in radon chamber for a term of 1 month. A diagram setup of SSNTD technique is represented in **Figure 7**. After that the CR-39 sheets have been etched in 6.25 normal NaOH solutions at 70°C .ffor 6 h. The etched samples were washed in running water and dried in air. The track density in each sheet is calculated to be n_1 and n_2 in the case of open and closed conical flasks, respectively. The radon permeability can be measured using Eq. (11).

Sample	Thickness (cm)	Track technique	Arafa method (2002)	Proposed method
LLD[b]	0.00250	1.27E−07	0.49E−07	1.5E−07
LD[a]	0.00245	1.2E−07	0.57E−07	0.9E−07
HD[c]	0.00295	1.1E−07	0.46E−07	0.46E−07

[a]LD: Low-density polyethylene
[b]LLD: Linear low-density polyethylene
[c]HD: High density polyethylene

Table 3. Radon permeability (cm^2/s) of polyethylene membrane using three different methods.

Table 3 shows the radon permeability (cm^2/s) of polyethylene membrane using three different methods for high-density polyethylene samples, and the permeability of radon is stabled fined by our offered method as the same value measured by Arafa system [61] for great density polyethylene samples. For low-density polyethylene sample, there is substantial covenant between the value of radon permeability constant determined by our offered method and that of SSNT technique.

6. Theoretical calculation of the calibration factor

The calibration factor of radon dosimeter (CR-39 track detector) for radon and its daughters has been determined theoretically. Calibration factor (K) is the quantity, which is used for converting the observed track density rates to activity concentrations of the species of interest.

If ρ is the track densities observed on a SSNTD due to exposure in a given mode to a concentration C for a time t, then:

$$\rho = KCt \tag{18}$$

The signal measured by etched track dosimeters is the integrated track density (tracks cm^{-2}). If the track density is uniform over the detector surface, the signal is defined by: [62]

$$\rho = \frac{K}{\tau_r}\int_0^{t_e} C(t)dt \tag{19}$$

where τ_r is the radon decay time (5.5 days), $C(t)$ is the radon concentration in the air around the detector (atom m^{-3}) at time t, and t_e is the exposure time, K denotes the dosimeter response (calibration coefficient) and is defined by:

$$K = \frac{\tau_r}{C(t)}\frac{d\rho}{dt} = \frac{1}{A_0(t)}\frac{d\rho}{dt} \quad (20)$$

where A_0 (t) is the activity concentration of radon in air (Bq m^{-3}) at time t and $d\rho/dt$ is the track density production rate (tracks cm^{-2} d^{-1}).

For a diffusion chamber-type dosimeter with a long exposure time (t_e >> .3 h), the response can be calculated by: [36]

$$K = \frac{\rho}{\overline{A}_0 t_e} \quad (21)$$

where $\overline{A}_0$ denotes the average radon activity concentration and t_e is the exposure time. In general, the total track density $\rho(r)$ at the point of observation in the detector, defined by r, is the result of an integration of all α-emissions which can produce observable tracks. It can be written as:

$$\rho(\underset{r}{\rightarrow}) = \sum_i \rho_i^v(\underset{r}{\rightarrow}) + \sum_i \rho_i^p(\underset{r}{\rightarrow}) \quad (22)$$

where $\sum_i \rho_i^v(\underset{r}{\rightarrow})$: track density due to radon, *Po*: isotopes existing in the air, and $\rho_i^p(\underset{r}{\rightarrow})$: track density due to radon decay progenies.

Ilic reported that for every α-produced nuclide existing the air is calculated by [63]:

$$\rho_i^v(\underset{r}{\rightarrow}) = \int_0^{t_e} dt \int_{V_1} \frac{A_i(\underset{r}{\rightarrow}, t)\cos\theta}{4\pi|\underset{r}{\rightarrow} - \underset{r'}{\rightarrow}|^2} dV \quad (23)$$

where $A_i(\underset{r}{\rightarrow}, t)$: activity concentration of ith nuclide for time t and at a point r, θ: angle between the particle's trajectory and the normal to the detector surface, $|\underset{r}{\rightarrow} - \underset{r'}{\rightarrow}|$: distance between the detector surface and the source volume dV, and V_i: the sensitive to volume for ith nuclide.

The stable rate of etching is calculated by the critical angle of etching. $\theta_{c,i}$ in this event can be found as:

$$\rho_i^v = 0.25 A_i t_e R_i cos^2\theta_{c,i} \quad (24)$$

Eq. (23) gives the estimation of track density as a function of the radon activity and intrinsic detector efficiency [64]. K_0 can be calculated by the flowing relation:

$$K_0 = 0.25 R_0 cos^2\theta = 0.88 \text{ cm} = 3.2 \frac{\text{tracks cm}^{-2}}{KBqm^{-3}\,h} \quad (25)$$

7. Future of radiation detection

In the last few years, radiation detection has emerged as an area of interest for quantum dots (QDs) application [65]. However, there have been very few published studies on the radiation detection based on colloidal QDs. The emission of QDs is size dependent, so the output wavelength can be tuned to the sensitivity of the PMTs or avalanche photodiodes (APD). However, the stopping power of most QDs is low and their scintillation luminescence is very weak [66]. The combination of high stopping power of inorganic scintillator with QDs could potentially lead to a new class of scintillator [67]. The aim of our future work is to develop the sensitivity and counting efficiency of radiation detector using nanoparticle.

Acknowledgements

This project was funded by King Abdulaziz City for Science and Technology, the kingdom of Saudi Arabia, grant number: 14-ENV1162-60. Financial support by King Abdulaziz City for Science and Technology is gratefully acknowledged.

Author details

Ayman M. Abdalla[1,2]*, Ahmed M. Ismail[3] and Tayseer I. Al-Naggar[1,4]

*Address all correspondence to: aymanabdalla62@hotmail.com

1 Department of Physics, College of Science and Arts, Najran University, Najran, Saudi Arabia

2 Promising Centre for Sensors and Electronic Devices (PCSED), Najran University, Najran, Saudi Arabia

3 Physics Department, Faculty of Education, Ain Shams University, Roxy, Cairo, Egypt

4 Department of Physics, College of Women for Art, Science and Education, Ain Shams University, Cairo, Egypt

References

[1] Durrani SA, Ilic R, editors. Handbook of Radon Measurements by Etched Track Detectors: Applications in Radiation Protection, Earth Science and the Environment. World Scientific Publishing; 1997. p. 387. ISBN: 9810226667

[2] Hararah SMAI. Investigation of radon pollution in groundwater in the southern part of Gaza Strip—Palestine [MSc thesis]. Gaza, Palestine: The Islamic University; 2007

[3] Human Health Fact Sheet. EVS, Argonne National Laboratory; August 2005. https://www.remm.nlm.gov/ANL_ContaminantFactSheets_All_070418.pdf

[4] Fleisher RL, Mogeo-Campero A. Mapping of integrated radon emanation for detection of long-distance migration of gases within the earth: Techniques and principles. Journal of Geophysical Research. 1978;**83**:3539–3549

[5] US Atomic Energy Commission. Personal dosimeter for radon and daughter products. U.S. Patent No. 749. 725. 680. 802;1968

[6] Cassou RM, Benton EV. Properties and applications of CR-39 polymeric nuclear track detector. Nuclear Track Detection.1978;**2**:173–179

[7] Tommasino L. Radon monitoring by alpha track detection. In: Tommasino L, et al., editors. Proceedings of the International Workshop on Radon Monitoring in Radioprotection, Environmental Radioactivity and Earth Science;1981; Trieste. Singapore: World Scientific; 1990. pp. 123–132

[8] Yoder RC. Radon gas detector with sensing film. U.S. Patent No. 211, 516, 880, 624;1988

[9] Wheeler RV. Detector for radon gas and its decay products. U.S. Patent No. 012; 266; 1987

[10] Pretzsch G, Borner E, Lehmann R, Sarenio O. Radon diffusion chamber with passive integrating detector. German (former Democratic Republic) Patent No. 276, 690, 850, 528; 1985

[11] Durrani SA, Ilic R, editors. Handbook of Radon Measurements by Etched Track Detectors: Applications in Radiation Protection, Earth Science and the Environment. World Scientific Publishing; 1997. p. 387. ISBN: 9810226667

[12] Kotrappa P, Dempsey JC, Hickey JR, Stieff LR. An electret passive environmental ^{222}Rn monitor based on ionization measurement. Health Physics. 1988;**54**(1):47–56

[13] Marvin HB. How to measure radiation with electrets. Nucleonics. 1955;**13**:82–87

[14] Kotrappa P, Dua SK, Pimpale NS, Gupta PG, Nambi KSV, Bhagwat AM, Soman SD. Passive measurement of radon and thoron using TLD or SSNTD on electrets. Health Physics. 1982;**43**(3):399–404

[15] Kotrappa P, Dua SK, Gupta PC, Pimpale NS, Khan AH. Measurements of potential alpha energy concentration of radon and thoron daughter using an electret dosimeter. Radiation Protection Dosimetry. 1984;**5**:49–51

[16] Gupta PC, Kotrappa P, Dua SK. Electret personal dosimeter. Radiation Protection Dosimetry. 1985;**11**:107

[17] Ramsey PM, Stieff LR. Programmable control exposure radon measurement system. U.S. Patent No. 920: 421; 1992

[18] Gast H. Measuring chamber for determining alpha activity of radon gas in carrier gas especially air. German Patent No. 200: 308, 920, 109; 1992

[19] Megumi K, Mjamuro T. Method of measuring radon and thoron exhalation from the ground. Journal of Geophysical Research.1977;**77**(17):3052–3056

[20] Perlman D. Method and apparatus for radon detection. European Patent No. 359: 770, 900, 328; 1990

[21] Horowitz YS. TL Dosimetry. Boca Raton, FL, USA: CRC Press; 1984

[22] McKeever S. Thermoluminescence of Solids. Cambridge: Cambridge University Press; 1985

[23] Hochman MB, Ypma PJ. Location of oil and gas reservoirs. Australian Patent No.01: 988, 788, 000; 1988

[24] Harley NH, Maiello ML. Passive measurement of environmental radioactivity. U.S. Patent No. 021, 450, 870, 304; 1987

[25] Seidel JL. Thermoluminescent dosimeter for monitoring subsurface radiation. U.S. Patent No. 941, 995, 780, 913; 1982

[26] Kotlyarov AA, Krivashcheev SV. Measuring chamber of volume activity of radon surrounding air samples. Russian Patent No. 2, 008, 694, 940, 228; 1994

[27] Matthies D. Radon concentration measuring instrument. German Patent No.940, 908, from 93DE-307, 015, 930, 305; 1994

[28] Thomson I. Radon detector. U.S. Patent No.4, 983, 843, 910, 108; 1991

[29] Dwcuir J. Radon monitoring device. European Patent No. 357, 697, 900, 314; 1990

[30] Iimasti V. Device for measuring radon gas concentration. Finnish Patent No.002, 717, 870, 618; 1987

[31] US Secretary of Interior Monitoring radon gas in air. U.S. Patent No. 653, 314, 760, 129; 1976

[32] Tirmarche M, Harrison JD, Laurier D, Paquet F, Blanchardon E, Marsh JW. Lung Cancer Risk from Radon and Progeny and Statement on Radon CRP Publication 115 Ann. ICRP **40**(1):2010

[33] Environment Protection Agency of USA. A Citizen's Guide to Radon. 4th ed. 2002

[34] Imtiaz N, Faheem M. Determination of 238 U contents in ore samples using CR-39-based radon dosimeter—disequilibrium case Radiation Measurements **41**(4):471-476

[35] Matiullah J. Determination of the calibration factor for CR-39 based indoor radon detector. Journal of Radioanalytical and Nuclear Chemistry. 2013;**298**(1):369–373

[36] Sutej T, Ilic R, Najzer M. Response of track-etch dosimeters to environmental radon. Nuclear Tracks and Radiation Measurements. 1988;**15**(45):547–550

[37] UNSCEAR. United Nations Scientific Committee on the Effects of Atomic Radiation Sources, Effects and Risks of Ionizing Radiation. New York: United Nations; 2000

[38] Dorsche lB, Piesch E. Radiation Protection Dosimetry. 1993;**48**(2):145

[39] Jamil K, Rehman F-U, Ali S, Khan HA. Nuclear Instruments and Methods in Physics Research Section A.1997;**388**:267–272

[40] Planinic J, Rdolic V, Faj Z, Suvelajak B. Nuclear Instruments and Methods in Physics Research Section A. 1997;**396**:414–417

[41] Yu KN, Leung SYY, Nikezic D, Leung JKC. Radiation Measurements. 2008;**43**:S357–S363

[42] Faj Z, Planinc J. Radiation Protection Dosimetry. 1991;**35**:265–268

[43] Al-Azmi D. Radiation Measurements. 2009;**44**:306–310

[44] Noorddin I. Natural activities of ^{238}U, ^{232}Th and ^{40}K in building materials. Journal of Environmental Radioactivity. 1999;**43**:255–258

[45] Wafaa A. Journal of Environmental Radioactivity. 2004;**75**:315–327

[46] Pylon Electronics Inc. Models110A&300A. Instruction Manual Number A900071;2009. p. 12

[47] Pylon Electronics Inc. Model AB-5. Instruction Manual Number A900024 Rev. 6; 2002

[48] DeCicco F, Pugliese M, Roca V, Sabbarese C. Applied Radiation and Isotopes. 2013;**78**:18–112

[49] Ashry AH, Abou-Leila M, Abdalla AM. Radiation Measurements. 2011;**46**:149–152

[50] Farid SM. Journal of Environmental Radioactivity. 1997;**34**:29–36

[51] Mansy M, Sharaf MA, Eissa HM, El-Kamees SU, Abo-Elmag M. Radiation Measurements. 2006;**41**:222

[52] Chu T-C, Liu H-L. Applied Radiation and Isotopes. 1996;**47**:543–550

[53] Singh K, Singh M, Singh S, Sahota HS. Radiation Measurements. 2005;**39**:213

[54] Tanner AB. Radon migration in the ground: A Review. In: Adams JAS, Lowder WL, editors. The Natural Radiation Environment. Chicago: University of Chicago Press; 1964;161–190

[55] Tanner AB. Radon migration in the ground: A supplementary review. In: Gesell TF, Lowder WM, editors. The Natural Radiation Environment III, Symposium Proceedings, Houston, Texas, April 1978, Vol. 1. U.S. Dep. Environ., Natl. Tech. Info. Serv., Washington, DC, Rep. CONF-780422, 1980

[56] Ramachandran TV, Lalit BY, Mishra UC. Measurement of radon permeability through some membranes. International Journal of Radiation Applications and Instrumentation. Part D. Nuclear Tracks and Radiation Measurements.1987;**13**(1):81–84

[57] El-Sammane H, Arafa W, Abdalla A. Temperature and humidity consideration for calculating air born ^{222}Rn using activated charcoal canisters. Health Physics. 2002:97–105

[58] Cohen BL, Cohen ES. Theory and practice of radon monitoring with charcoal adsorption. Health Physics. 1983;**45**:501–508

[59] Cohen BL, Nason R. A diffusion barrier charcoal adsorption collector for measuring radon concentrations in indoor air. Health Physics. 1986;**50**:457–463

[60] Montero Cabrera ME, Sujo LC, Villalba L, Peinado JS, Jiménez AC, Miranda AL, Herrera Peraza EF. Calibration of diffusion barrier charcoal detectors using a semi-empirical expression. Applied Radiation and Isotopes. 2003;**59**:281–287

[61] Arafa W. Permeability of radon-222 through some materials. Radiation Measurements. 2002;**35**:207–211

[62] Fleischer RL, Giard WR, Mogro-Compero A. Health Physics.1980;**39**:957

[63] Ilic R. Environmental Radioactivity and Earth Sciences. 1990;**133**

[64] Fleischer RL, Mogro-Compero A. Journal of Geophysical Research. 1978;**83**:3539

[65] Hossu M, Liu Z, et al. Applied Physics Letters. 2012;**100**(1):013109

[66] Yao M, Zhang X, et al. Journal of Applied Physics. 2010;**108**(10):103104

[67] Sahi S, Chen W. Radiation Measurements. 2013;**59**:139–143